THE COMPLETE GUIDE TO

PATCH BIRDING

HELM
Bloomsbury Publishing Plc
50 Bedford Square, London, WC1B 3DP, UK
Bloomsbury Publishing Ireland Limited,
29 Earlsfort Terrace, Dublin 2, D02 AY28, Ireland

BLOOMSBURY, HELM and the Helm logo
are trademarks of Bloomsbury Publishing Plc

First published in the United Kingdom 2026

A catalogue record for this book is available from the British Library
Library of Congress Cataloguing-in-Publication data has been applied for

ISBN: PB: 978-1-3994-1969-7;
ePub: 978-1-3994-1970-3;
ePDF: 978-1-3994-1967-3

2 4 6 8 10 9 7 5 3 1

Designed by Dominic Robson
Maps on page 27 and 45 by Julian Baker

Printed and bound in Dubai
by Oriental Press

THE COMPLETE GUIDE TO

PATCH BIRDING

Making the most of your local area

Ed Stubbs

HELM

LONDON • OXFORD • NEW YORK • NEW DELHI • SYDNEY

Contents

Preface

Birding means different things to different people. For many, it's a form of escapism – a pastime that takes one away from the hustle, bustle and hyper-connectivity of modern-day life. Of birding's many forms and niches, patch birding – watching a certain site or area, generally close to home – perhaps reflects this the most. It's about being at one with a place and a period of time, and the birds within that. It's impossible, then, not to salute the various areas that have inspired me over the years, since I was a young child, and nod to them following the creation of this book.

I was first introduced to the concept of patch birding by an assistant teacher at my infant school – Mr Marsh (not a bad name for a birder). He himself was new to birding but very keen and, after realising my enthusiasm at such a young age, he took me, accompanied by my mum, around our local sewage farm one grey September morning. Not every kid's idea of a good time, and I'm sure my mum was a little perturbed at the location, but I was hooked. A splendid drake Gadwall – the first I'd ever seen – among a flock of Mallard stood out the most. I still have the notebook I wrote in upon my return, complete with a stain next to the entry for that day from a hot chocolate that I remember sipping while eagerly jotting down my observations.

For several years afterwards, I visited that sewage farm as frequently as possible, largely on weekends when my parents could take me, and I keenly documented everything I saw. I was captivated by the stories the regular birders had of rarities in years gone by, religiously thumbed through the wire-bound annual reports they produced and was particularly fascinated by the 'main man', Brian, and his extraordinary dedication to the site.

Opposite page: One of my earliest memories of birding in my local area is seeing my first-ever Gadwall in a large flock of Mallard.

Time went on, and my interest levels waxed and waned during my teenage years. I went to university and returned home, and the bug was back. In all that time, no other form of birding – or indeed natural history – has captivated me quite like patch birding has.

There are plenty of people to thank. Mr Marsh for sharing his own enthusiasm. My parents for encouraging my passion from a young age. Brian, the sewage-farm patch superman, for demonstrating commitment to local birding and various words of wisdom shared when I was young. I can't forget my late Nana, either, whose help identifying a Redpoll in our garden when I was very small probably provided my birding 'trigger' moment. There are friends, too, some of whom kindly contributed to this book, including Amy, Dave, David, Matt, Josh and Josephine, who I regularly converse with about the highs and lows of patch birding. All of them – and many others – have provided inspiration and encouragement. And, of course, my partner Nadia, who has shown patience and understanding that has allowed me to indulge in my beloved hobby for a long time.

Finally, though, and as cited at the start of this segment, I must thank my patches – after all, they truly made this book possible. From unassuming arable fields and small hills to rubber reservoirs and soggy river meadows, the successful panning for avian gold they have hosted over the years ultimately encouraged me to write this book the most.

Kingfisher

CHAPTER ONE

The benefits of patch birding

The birds you encounter close to home are the perpetual characters of your daily birding life. You may even come to know them as individuals, as you grow familiar with the rhythms of their daily existence and annual cycle. Whether it's casual garden observations or time spent birding at a nearby nature reserve, observing how things change through the seasons is one of the great joys of being a birder. Sharing a close affiliation with a site or area makes us appreciate it – and the birds it holds – even more. Patch watching hones our birding skills, adds to our knowledge and understanding of birds' lives and provides important data, especially in lesser-watched regions.

The joys of a patch

Patch watching is a well-established part of birding culture. A local patch is, in simple terms, a site or area close to where you live that you visit on a regular basis to go birding. There are no set rules as to what a patch should be exactly – it could be a small park five minutes away or an area the size of a small borough that incorporates a variety of different habitats. The key is that it is local to you and somewhere you visit regularly. Furthermore, you can put in as much or as little effort as you like. You could visit your patch once a week, or you could visit it multiple times a day – if you regularly watch it, it's your patch. Your records of early or late arrival and departure dates for migrants, high counts of particular species, breeding records of scarcer species or sightings of rare visitors all help to build a better picture of an area's avifauna. Appreciating the birds close to home is undeniably a source of great satisfaction.

Dedicated patch birders have made up the fabric of British birding for decades. More casual birders can't help but respect anyone who can dedicate significant time and effort to a single area, though may also suppose that there must be some slight madness – or at least bloody-mindedness – going on for a birder to be so devoted to one place!

Patch birding gets you outdoors and amid nature, encourages exercise and puts you in a position to enjoy birding on your doorstep. This headland on the Dorset coast offers both migration excitement and splendid views.

The notion of studying wildlife in a local area is deeply engrained in British birding culture. You can go back as far as the eighteenth century and Gilbert White's seminal *The Natural History and Antiquities of Selborne* to find what are arguably the first documented patch studies, with White's charming notes on the wildlife year in his part of rural Hampshire an early example of the ardent study of nature in a local area. As birding developed in the UK through the twentieth and twenty-first centuries, the notion of patch watching grew and it swiftly became a popular part of the hobby. Indeed, much of the vast dataset on British birds has been garnered from years of dedicated patch study via citizen science (just take a look at the RSPB's long-standing Big Garden Birdwatch). Many of the rarest birds you've seen will have been found by patch birders. In many ways, patch watching is the very lifeblood of British birding.

I have long thought that birding can be split into different subcultures. Particularly these days, there are so many different components to the hobby and a range of ways to engage with it that to call yourself simply a birder is quite broad. You have patch watching, twitching (travelling to see individual rare birds found by other people), listing (adding to the list of species seen in a year or in your life), photography, rarity-finding, migration study (including vis-mig), global travel, ringing, sound recording and so on... all quite different types of birding, and all enjoyable in their own right. Some people partake in only one or two of these birding 'genres'; others, myself included, indulge in several.

I have been on thrilling twitches and birded in some extraordinary parts of the planet. I've also witnessed some of the world's most spectacular migrations, and get a kick out of photography, too. But none of the buzzes that each has provided has come close to the feeling of a good day or a good bird on my patch. Connecting with a mega-rarity that you've twitched is exhilarating – but for me a twitch comes with an added dose of stress that patch birding rarely kicks up. As a hobby, I believe birding is about moments and for many a long-desired tick or top-quality photograph might provide that. But ultimately, that chase for a new bird to add to your 'life list' or uploading an image on social media is for a single end-result – patch birding is about moments that will linger longer in the memory, and it provides endless potential. Encountering unusual species on your doorstep or watching the avian year play out in a certain area creates a deeper connection with a period of time and location that other elements of the hobby can't provide in the same way.

My very first birding memories are either of garden encounters or sightings during walks close to home. That early connection with local birdlife has continued throughout my birding career and patch watching has been the bread and butter of my avian intake ever since. I believe that the thrills of local birding can be enjoyed by anyone who tries, even if you live in what may be perceived as an uninspiring birding region or are a beginner navigating your way through the hobby for the first time. I also believe that every patch birder – even those with decades of experience – can improve their skillset and get even more satisfaction out of their efforts. This book aims to help both the novice and the more experienced birder.

Patch vs local

At this point, it's important to touch on the idea of local birding versus patch birding. They fall under the same umbrella – this book is about both local and patch birding – but they can take on different meanings. While a patch generally refers to a single site or place, local birding tends to mean a wider area – perhaps a certain radius from your house or the parish or borough that you live in. In recent times, there has been something of a shift in Britain from single-site patch birding to local-area birding. The reasons are varied and, among others, include things like the restriction of movement during the COVID-19 pandemic in 2020 and 2021. The various levels of lockdown forced people to stay close to home, resulting in their birding world shrinking to their immediate garden or local area. The heightening climate crisis has led many to revise their birding approach as well. Sticking close to home and undertaking a greener, low-carbon version of the hobby is commendable, and birders doing this often like to expand their birding options by covering a wider area instead of a single locale. It is possible of course to have patches within a local area, too. Below is an outline of the

Understanding what habitats are in your local area is key to knowing where the birds will be.

meaning of some important phrases that'll crop up throughout this book:

Patch A single local site that you bird regularly. This may be an urban park or gravel pit, or a coastal headland or estuary, for example. It will have a clear boundary, established either by the patch birder themselves or through geographic features. It is easily covered on foot during a single visit. In rare cases a 'mega patch' may be a contiguous recording area, which can be the case for larger, well-established birding sites (Spurn in East Yorkshire and Dungeness in Kent, for example).

Local area A local region in which you bird regularly. There are various ways of defining your local area – a parish or borough boundary, or a certain radius from your front door, for example. Within a local area you may have multiple patches (and seasonal sites and new ground, see below).

Seasonal site A site within your local area that you bird semi-regularly or more at certain times of the year than others. The difference between a seasonal site and a patch is the consistency of coverage (which is outlined in the patch guidance criteria below).

1km The area within a 1km radius of your home. This is used by some birders for 'micro-patching' – birding the immediate surrounds of where they live.

New ground Parts of your local area that you rarely cover or have yet to. These make up an exciting and fun part of birding in your local area and are covered on page 112.

Patch rules

As touched on earlier, there are no formal rules or regulations as to what can constitute a local patch. However, it can help to have some guidelines. So, below is a list of suggested criteria for helping define what a patch is:

- A patch must be within a 10km radius of where you live and a local area must be no larger than a 10km radius of where you live. The closer to home, the better.
- A patch, or the most-distant part of a local area, can't be longer than a 20-minute car journey from where you live. Ideally, you can reach a patch or cover a local area by green means (on foot, by bike or by public transport, for example).
- A patch or local area must be birded regularly. It's hard to define 'regularly' as the amount of available birding time varies from person to person. Ideally, though, a patch should be birded at least once a month year-round, and a local area should be birded once a week year-round.

These criteria are, of course, subjective. Depending on where you live, the habitats around you, how often you can go birding and how easy it is for you to get around, what constitutes a patch or local area will vary. You might have a work patch that you visit on lunchbreaks – and where you work may be far from home. There is also the concept of 'green birding', which basically means that anywhere you can reach on foot or bike counts, and if you're particularly fit and mobile this can comprise a huge area. It's ultimately up to you to define your idea of local.

Why local birding?

There's no doubt that, to some, the idea of patch and local birding is not appealing. For starters, more often than not, you have to work to find your own birds. It is not a simple case of rocking up and seeing something unusual or exciting, as you might on a twitch or a foreign trip. Furthermore, venturing beyond your local area may take you to birding sites that are productive and stimulating – why would you bother traipsing around close to home when you can go further afield and be guaranteed birds? And how can you build a good life or year list if you're spending most of your time locally, where bird diversity might appear lower?

Well, there are a multitude of reasons why patch and local birding is the way to go. By watching a certain site or area over a period of time, you come to feel that you're part of a time and a place. You notice the little details of the natural world as it runs through the seasons – a sense of immersion and shared experience that is truly special. This runs through to watching birds with far greater purpose and focus. I'd also suggest that your birding effectiveness develops, too; patch watching undoubtedly forges improved levels of attention to detail.

As you visit your local patch or bird in your local area more, you will inevitably build up a knowledge of the local birds – not just in terms of identifying them, but also their behaviour and habits. You will paint a detailed picture of both the resident and visiting birds, and notice subtle changes or oddities that would otherwise go unnoticed. Throughout the year, with each passing month, you'll recognise the understated changes of the birding year – not just your first Swallow of the spring, or Redwing of the autumn, but the first singing Greenfinch of the year, northbound Black-headed Gulls and nest-sitting Eurasian Coots. Identifying these subtle seasonal shifts is

Noting the first time you hear familiar birds, such as Greenfinch, singing each spring is one of the joys of patch birding.

one of the most appealing things about local birding, tying you in with the flow of the avian year in a way that almost makes you feel like you're a part of it.

Over time you will get to know which areas of your patch are best for certain birds, too – and you'll be able to implement specific tactics to try and locate scarcer species. That fence-line that attracted an early spring Wheatear this year may do so again, so it's worth checking when that time of year comes around once more. And that waterbody that remained unfrozen during a cold snap, attracting wildfowl, is worth inspecting during the next chilly period.

Frequent coverage means you'll notice when something unusual is present. Migration plays a huge part in the excitement of patch birding and it can be experienced almost anywhere, from the middle of big cities to your garden (should you be lucky enough to have one). Migration can bring surprises and, if you're tuned in to your local area and what birds are to be expected, you are more likely to notice when something unusual passes through. Furthermore, your knowledge of where the common birds are will help to give you the best chance of locating a scarcity.

As time goes on, you will build up a broad understanding of your local area and the sites and habitats within it. Simultaneously, your 'patch portfolio' will grow, with an increasing range of rare finds, important breeding data and, most importantly, birding memories under your belt.

Going green

A harder-hitting reason why local birding appeals is the ongoing climate breakdown – a huge, scary reality that is obvious to the observant patch birder. Put simply, a reduced carbon footprint has clear benefits for the health of our planet. We must all look for ways to lower our carbon footprint, and a greater focus on local birding is a fine way of doing just that, especially if you can undertake much it on foot or by bike. Even if you're simply driving less, it's a step in the right direction.

I touched on birding subcultures earlier and some of these are closely entwined with heavy mileage – be it national listing, world birding or even relatively casual regional year listing. Breaking these traditions is not easy, and I will hold my hand up as a birder who has and does indulge in 'longer-distance birding'. But the reality is that we all need to try and reduce these activities to some extent, even if cutting the cord seems difficult.

Local birding means travelling shorter distances, perhaps also by greener means. Local birding also means, by default, more time birding – less travel time equals more field time. Your chances of bumping into something unusual are raised when you're outdoors, too – you might not have heard that singing Nightingale in a roadside thicket if you were driving in your car with the windows up, but if you're on your bike you're plugged into nature. I can think of plenty of decent birds I've found while 'in transit' on patch, located in inauspicious areas between more favourable sites.

The feeling of satisfaction at the end of an enjoyable session on patch is hard to match. You will certainly feel like patting yourself on the back if you plumped for some local birding and it was productive, instead of getting in the car and driving further afield.

A bicycle is an excellent way to cover your local patch. It gets you around quickly, constitutes low-carbon travel and provides health benefits.

Citizen science and safeguarding birds

It's not hard to segue from the climate crisis to the decline of many of our birds. Plenty of species are suffering miserable declines and have been for decades. Some migrants are turning up earlier or later than usual, while some winter species are 'short-stopping' (essentially not bothering to visit Britain in the winter months due to the lack of cold spells on the continent).

In these changing times, the study of our birdlife has huge significance. Each birder's personal observations and records hold genuine importance – and meticulous field notes over a period of time can help create an understanding of the changing status of different species. Taking part in citizen science is exceedingly valuable, and it is also easier than ever to add your sightings to bigger regional and national pictures, with digital platforms like BirdTrack and eBird (see pages 56–58). Collectively, the data provided by local birders helps paint a picture of what is declining, what is increasing and so on. And recognising such trends is the first step towards understanding what is driving them – and ultimately the implementation of conservation schemes to safeguard those species that are struggling.

So, the birds in your local area can benefit from your presence, and it's not just about declines and losses. For example, imagine a scenario where your patch is being threatened by building development. By keeping (and sharing) your patch records, you may be able to prevent it from being built on or developed – it might be that your records reveal the presence of a rare or sensitive breeding bird. Birders – and indeed wildlife lovers in general – are in the minority in Britain, so it's important we stand up for nature and offer some form of voice in the face of seemingly ever-increasing development – and your records can help form that blockade.

Individual observations are so important and make up key data, which are often rounded up at county level by county bird clubs and societies. Given how simple it is to submit records – and how important it is – I struggle to understand why a birder wouldn't do so. I also believe that keeping records (and sharing them with county clubs and national and global bird recording platforms) improves your birding skills. To find out more, see pages 56–57.

Benefits beyond birds

Not only is birding on foot or bicycle better for the environment, it is also healthier. By ditching the car and getting out on foot or bike, you are going to do more steps and burn more calories. If you watch a local area, travelling between sites by walking or cycling is a great way to improve your fitness and indeed bigger patches may require a few miles of walking anyway.

Patch birding also offers an important form of escapism. If you've had a stressful day or are enduring a difficult period, picking up binoculars and heading outside can sometimes be the perfect tonic, enabling you to switch off and immerse yourself in your local birdlife. In this hectic modern age, simply being alone somewhere quiet provides welcome respite. Being outside is well-documented as being a great way to improve your mental health, and local birding offers this in abundance.

Staying local can make a positive impact on your wallet, too. Maintaining a car and travelling around are increasingly expensive. Twitching, especially long-distance, requires hard cash. Even driving to and from a site an hour away is going to add up if done with any regularity. Patch watching is a cheaper, financially friendly way to bird. If you can get your hands on some budget binoculars and are able to get outdoors, then you're set.

The camaraderie that local birding forges is another huge positive. I have made lifelong friends from birders I first met in my local area. When there is a group of you watching a particular site or living in the same area, a sense of community and shared experiences is generated that it really adds to the whole patch picture, while also allowing you to learn from others – and encourage beginners or birders new to the area.

Love local

Of course there are downsides to local birding, as with any form of birding. There can be moments of frustration and disappointment, as well as occasionally long periods of the doldrums. Overall, however, it is an enthralling, captivating way to spend time. And the little moments of magic that eventually do come along create unforgettable memories and generate

a unique buzz. For more on this, see pages 178–181.

Keeping detailed records enables you to look back and recall the exact date, time and even place that you saw different birds. On future visits to your local patch this means you can either check that same place for more rarities, or simply remember the past encounter and smile.

It's important to stress that patch and local birding needn't be your only form of birding. Variety is indeed the spice of life and it's good to mix it up – this will also help to keep your birding sharpness and maintain your hunger to get in the field locally. But I am confident in saying no other form of birding – however great ticking that bucket-list world bird or hitting a landmark for your British life list feels – provides as much joy as watching a patch or local area.

About the book

The rest of the book will aim to provide all the information you need – whether you're new to birding or experienced – to get the most out of birding in your local area or on your patch. I'll begin with how to pick or outline your patch – where to patch, what a good patch needs, and how to identify one.

I will then discuss a massively important element of patch birding – time and weather. What time of day – and year – is best? How long should you visit a site for? What weather do you need to look out for? Knowing these trade secrets is key to being a productive local birder. A look at getting the most out of your patch results and contributing as best you can will also be covered.

A month-by-month run through the British birding year will offer a range of tips and tricks to use throughout the seasons. There is always something to seek out on patch, whatever the time of year. I will also cover the importance of record keeping, the mental strain that patch over-exertion can cause and the best kit – from binoculars to apps – for the active patch-watcher.

During my long time as a patch and local birder, I have taken on board advice and tips from various peers. Understanding weather, time of day, the rhythm of the seasons and everything else is very important when it comes to maximising your patch time, and much of that is demonstrated and taught by others. However, it is often the tales – gripping yarns – of other patch birders that provide the deepest inspiration and encouragement. They can serve as fuel to get in the field and try to get lucky in creating your own special moments. Every patch birder has their own journey and stimulus too, and it's important to remember that not everyone is blessed with a brilliant patch or that seemingly magic knack of routinely finding good birds. 'Patch tales', interspersed between chapters, feature contributions from five patch birders, all hooked by this element of birding. Hopefully their words will provide insight and encouragement.

Patch and local birding is fulfilling and wonderful. With luck, some of the information provided in this book, or some of the stories told, will encourage you to get out locally more often or work your beloved patch that bit harder.

Left: Over time studying a certain place in your local area, you'll begin to know every footpath, field and copse well, each with their own birding stories from previous visits.

Tufted Duck

CHAPTER TWO

Picking a patch or local area

So, you want to find a brand new birding patch to make your own? Perhaps you're new to this form of birding, or have recently moved to a new area. Or maybe you're a long-time patch watcher who wants to expand your selection of local sites. Where do you start when picking a patch – what elements does a good patch or local area need to help you get the most out of your time in the field?

What makes a good patch?

A local patch is, typically, somewhere close to where you live that you can visit on a regular basis and get to know the birds that reside there. It can be anything – a local park, a small area of woodland, a lake, a gravel pit, an estuary, a stretch of coastline or even the top of a mountain. The only real rule is that it needs to be local to you. See page 13 for what constitutes local. In terms of size, it's worth considering the distance you can regularly cover in a timeframe that fits your lifestyle.

When identifying a patch, you need to consider access. After all, there's not much point in having an excellent patch half an hour way that you can only visit every other week. A local patch should be somewhere that offers easy access so that you can visit regularly, at least once a week at various times during the day right throughout the year.

For some, choosing a local patch is easy, as they may live within close range of a variety of superb bird habitats, perhaps including actual nature reserves. For those lucky few, the only problem they may face is perhaps the presence of lots of other birders. If, however, you fancy a challenge, why not be a pioneer and try somewhere off the beaten track – this way of local birding can be particularly rewarding, after all.

Where shall I patch?

An initial decision you need to make is whether to pick a single patch to cover, or to work a larger area. Each option has pros and cons, outlined in the table opposite.

Strong cases can be made for both. A lot of it boils down to what your local area offers and what the trade-offs would be. If, for example, you live on the east coast next to a wetland, then

Patch	
Pros	Regularly birding the same site allows you to understand it on the deepest level, maximising your knowledge of certain areas or seasons.
	You are reducing time travelled between sites while increasing time in the field.
	By being in one place a lot, you increase your chances of finding unusual birds – a factor to consider if rarity-finding makes you tick.
Cons	Repetition can lead to lessened concentration, resulting in diminished scrutiny of certain spots or gatherings of common species. It can also lead to frustration.
	Your typical ensemble of birds stays largely the same, and the possibility of discovering new sites for rarer species is reduced.
Local area	
Pros	Multiple sites = multiple opportunities to study birds, with different habitats offering a different suite of species.
	You can keep things fresh by visiting a range of sites.
	By covering a larger area, you will gain a greater understanding of the birdlife in your particular region.
Cons	Travel time is greater, with a bigger area to cover.
	Time on your better patches within your area is reduced, as you're not visiting the same site every time, which could mean you're missing visiting birds when you aren't at said site.

it would make sense to patch that site alone. On the other hand, if you're inland or in a city, with birding hotspots thin on the ground and further apart from one another, then collating several under the banner of a local area is a better way to get more out of your time in the field.

Two Ts are also important: time and transport. If you can only get in the field once or twice a week, then perhaps you're better off visiting the same single location as much as possible. If you're able to go birding more regularly though, perhaps even daily, then you can afford to 'spread the load' and cover a greater selection of sites.

Left: It's important to understand on what land you can and can't go birding.

How you get around will vary depending on where you live and your personal circumstances. In an ideal world, patch birding would be undertaken in the lowest-carbon form possible – as much on foot or bicycle as can be. This isn't realistic for everyone, but however you get to your patch or cover your local area, it needs to be relatively effort-free. A complicated bus journey or traffic jam-ridden 20-minute drive is not ideal. You want to maximise your time birding, so trying to find somewhere you can access regularly and easily is important.

I used to cover a single patch, but now watch a local area and this, I believe, is an increasingly popular trend among birders. I think the COVID-19 pandemic played a role in fast-tracking the 'local area' concept, with the restrictions on how far we could stray from our front doors shrinking people's worlds – and, from the point of view of a birder, greatly reducing the birding playground. Naturally, people wanted to make the most of their area as much as they could during these times, leading to all sorts of local birding year-listing competitions in 2020 and 2021 across the country, often within certain radiuses from home.

This, as well as the climate crisis, reinforced the idea that we need to learn to appreciate the places closer to home as travelling longer distances is not – or in the case of the pandemic was not – possible. For me, birding a local area is the clear winner. But that's because I live in landlocked Surrey where birding hotspots are sparse. This means that, by covering a certain area, I can incorporate a range of habitats and sites. As a result, different forms of birding are on the cards, be it scanning winter duck flocks, autumn vis-mig, seeking out rarer breeding species, or targeting wader passage. It might be a different scenario if a productive nature reserve such as Cley Marshes or Pagham Harbour were on your doorstep. So, work out what's best for you, both in terms of quality of birding and practicality.

Of course, there's no reason why you can't do a mixture of things, for example selecting two or three patches that are disjointed but within a 10km radius from home, or something similar. And it's worth mentioning 'secondary' patches too, which many people have – these might include a park near your office that you visit for lunch-break walks, for example.

I bird a local area – and within that I have a few patches. My local area is

This landscape might not look special, but some useful patch-birding habitats can be seen. These include a hedgerow, a fencefor perching, scattered trees and, in the distance, raised ground.

actually based on the recording areas of the Surrey Bird Club, which split the county into four different spaces a few decades ago. I live in the middle of one of them, south-west Surrey, and all of my regular patches are within it, too. So, it made a perfect local area when I first moved from watching a single patch. South-west Surrey's boundary is defined by borders with Hampshire and West Sussex in the west and south, the A281 in the east and A31 in the north. In terms of size, it's not dissimilar to a medium-sized borough. The furthest site from home is a 20-minute drive; most spots, however, are much closer and I do the majority of my birding within a 4km radius of my front door. This includes four or five sites that qualify as patches, such is the regularity with which I visit them compared with other places in south-west Surrey. There may be things like this that can work for you – check in with your local bird club.

I know plenty of people who use parish or borough boundaries for their local areas, too, and this is a really good way to settle on a pre-defined space in which to do your birding. The internet is your friend in this case – have a play around online to see if you can find any maps you may not have realised even existed.

What does a good patch need?

It's all well and good picking out your patch or local area – but what do they need to be the most rewarding possible? If you're dedicating yourself to patch birding, you'll want to be able to maximise your defined space as much as possible, squeezing out every last drop of possibility for good and exciting birding. If you're picking a patch or local area from scratch, then there are a few things to consider in order to get the most out of your time in the field.

A nice mix of habitats can be seen in this photo, including a river, a floodplain, woodland and grassy meadows, as well as plentiful sky.

Water is an important part of a patch. It doesn't need to be a large wetland - even a small rubber reservoir on a farm can attract birds.

Patch

If you're going down the single site route, then it's important that your patch offers either dynamism, in terms of habitat, or has good form for birds. Birders, especially those inland, often go for wetlands – and rightly so, but it can pay to pick a wetland that has other habitats adjacent. Scrubby meadows, a small block of woodland, arable fields or even some parkland all offer a different suite of birds to, say, simply a marsh or lake. Each adds a bit of variety to your patch watching, which is important during the inevitable quiet spells.

On the other hand, I know plenty of birders who relentlessly watch one-dimensional – but productive – sites. Large inland reservoirs are good examples of these. They offer little variety on a day-to-day basis, but will often have good form for producing rarities and spells of entertaining migration. This is all some birders want from local birding. So, if a single site near you has good form for unusual birds, and this is your primary motivation for getting in the field, then it's an option – but be prepared for repetition!

Somewhere with a bit of height (or merely some open sky) is a real boon for any patch (and indeed in any local area). The power of vis-mig is huge when it comes to patch watching, for it can produce brilliantly entertaining days of migration and rare (on a national, regional or local level) birds. So, try and find somewhere that offers an elevated vantage point.

Picking somewhere that doesn't suffer too much from disturbance is ideal, too, though this is increasingly difficult to do. In a perfect world, dogs and joggers won't have covered your patch before you have in the morning, but this isn't always realistic.

Local area

As with picking a patch, dynamism is important when picking your local area. The difference is that you can only work within a certain radius or set boundaries – but having a local area with a few half-decent sites can combine to be a productive one. The most important element is variety. So, try and ensure that your chosen local area includes at least one notable waterbody, whether it be a reservoir,

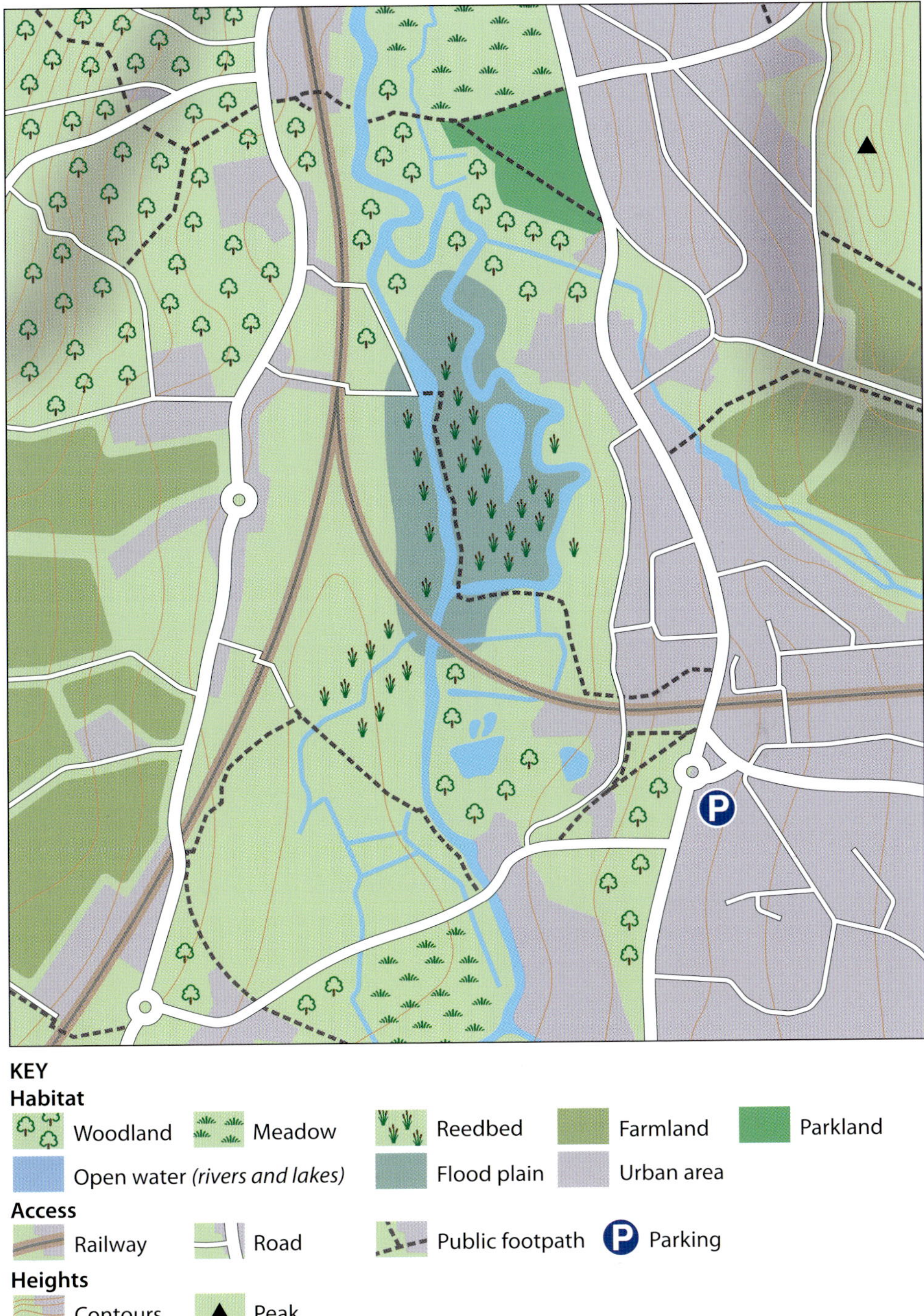

This is an example of a fictional area that would make for a good patch. It encompasses a variety of productive habitats within a manageable area and is easily accessible via roads and footpaths.

lake, pond, river or estuary – some form of water is the key. However, having good areas of woodland, farmland, meadows and scrub will greatly diversify your birding, allowing you to target and look for species that would be virtually unthinkable in other parts of your local area, depending on factors like time of year and weather. For example, I know I'm very unlikely to see Goshawk, Marsh Tit and Hawfinch at my local reservoir, but a woodland about a mile away does yield such birds. Once you get to know the different sites and habitats in your area, you will soon learn to maximise each one's birding potential.

That being said, you don't want to bite off more than you can chew. Are you totally reliant on walking? And how much can you get out – if you're barely going to visit most of your chosen area in an average year, then perhaps a smaller radius would be better? Sometimes if you cover a larger local area you can feel like you're not reaping the full potential of certain spots.

A well-covered local reserve can make a great patch and may come complete with trails, parking and even hides.

Another thing to consider is how well-watched a patch or area already is. A popular RSPB reserve, for example, may appeal to you – established trails, hides perhaps, and plenty of other visiting birders might be enticing. On the other hand, you might like the idea of chartering new ground, or simply going somewhere where you are unlikely to bump into other birders. It's all down to the individual, but it's worth considering if there are any honeypot sites in your area.

Summary

Whether it's a patch or a local area you're covering, look for:

- Habitat dynamism
- Relatively easy access and short distances to cover
- Areas that are less disturbed by other people
- A spot or two with some elevation or open sky

How do I find a patch?

Once you've decided whether you want to watch a patch or a local area, and you know what sort of features to look for in both, it's time to find it. Google Maps and OS maps are very helpful at this point. I find Google Maps to be particularly useful for finding stretches of river, little ponds or reservoirs, or blocks of promising-looking woodland or farmland. Google Maps also allows you to work out any distances on foot, or by bicycle or car. The same can be said for OS maps, which doesn't have the interactivity levels of Google Maps, but does show footpaths – crucially – and contour lines, which are useful for things like identifying promising vis-mig spots.

I've spent hours poring over maps, trying to find new areas to visit that may have patch potential. I also spent lots of time playing with different radiuses and boundaries before settling on my south-west Surrey local area. The more you search, the greater your chances of finding somewhere good. You could discover a little stretch of river valley and adjacent farmland and woodland that is criss-crossed with path. Or, by slightly shifting your, let's say, 5km radius in a certain direction, you might be able to include a promising-looking bit of heathland or marsh.

You can try before you buy, too. If you're new to an area, or new to birding or to the concept of local patch watching, then why not test out a couple of promising areas and see which you enjoy the most, or which offers what you deem to be the best potential and easiest access. Start off by making a few reconnaissance visits at different times of the day and make a note of the species that occur there. Make sure that you note down the number of individual birds too – the more birds (of whichever species) at a certain place, the better its potential. Keep an eye on the time that it takes you to cover the patch as well.

When I moved back from university, I wanted to pick a 'new' patch and had a choice of three slightly different sites, were a similar distance from home and were under-watched. I spent the best part of six months trying each out, before settling on one that became the focal point of my birding for the next five years, before expanding to my south-west Surrey local area.

In a similar vein, it can be handy to read up on a local area to better understand it. If your county has a published book on its avifauna, then try and get a copy. Reading through different species accounts can paint a historical picture of the productivity of a certain area. County bird reports

This river with adjacent vegetation and footpath offers access and habitat diversity.

Maps are handy if you're looking to patch somewhere new or explore different places in your local area.

offer similar information, as do the search functions of BirdGuides and eBird. You might conclude that a certain area seems to have been particularly productive down the years, even if it's a little off-piste or isn't a popular destination. Such locales can be excellent patches!

Most of us aren't lucky enough to live near prime bird habitat, so choices when it comes to a patch or decent local area are somewhat limited. If you're stuck slap-bang in the middle of the suburbs of a major city, for example, then the best thing to do is to look at a map of your area with your home as the focal point and then pinpoint potential local patches within easy walking or cycling distance. Look for 'islands' of habitat – with regular coverage you'll be surprised at what you can find.

Once you've decided on your patch or local area, it's important to define its boundary. This reinforces the idea of a location's identity, while also encouraging you to maximise a distinct space. If your patch is an official reserve or public space of some form, then it may well already have a defined boundary. If not, using the My Maps tool on Google Maps is a fun way to draw up a boundary. When doing so, consider the size of the site and how much coverage you can give it.

If going for a local area, then there are several options. Free online mapping programmes, such as mapdevelopers.com and freemaptools.com, are easy to use and can be good for working out radiuses in particular. A popular method for birding a local area is to use a circle radius from one's front door. Depending on where you live and the types of habitat you have, anything from 5–10km is probably optimum. Anything beyond 10km is probably a bit of a stretch for 'local', but it's ultimately down to what works best for you. It may be tempting to push a boundary to include a certain site or bit of habitat, but try to embrace the challenge of working within a limited area – that's what patch watching is ultimately about, after all!

Micro-patching

If you're covering a local area, but have a particular favourite long-term patch within it, there's nothing wrong with that being your primary focus for your birding and your default location for most outings, with other sites in the wider area more supplementary and secondary in nature. As well as that, it can be fun to discover new 'micro-patches' within your local area. For more on this, see pages 59–60.

Identifying new areas

This is a hugely important element of patch birding that is applicable whether you're vastly experienced with your local area or not. Outside of protected reserves and areas of land, habitats are often changing – normally for the worse, it must be said, with seemingly never-ending development in much of the British countryside. Despite that, new areas of patch promise can appear so keeping your eye in for such areas is advisable.

For example, our increasingly wet winters may produce new areas that experience regular flooding, especially along river valleys or on low-lying farmland. Birds are amazingly quick to identify new foraging and resting sites, and such areas can suddenly be filled with a variety of species. A small irrigation reservoir might be constructed on some farmland, with waders attracted during migration, or an area of woodland might be clear-felled, with Nightjars and Woodlarks moving in temporarily.

Keep your eye out for change and test new spots if they appear. Sometimes ephemeral places can produce remarkable birds. A record that stands out in my mind is that of Surrey's only ever Marsh Sandpiper, which was discovered on a small flooded field that had come about after a sewage works pipe had burst. The little wetland was in existence for mere weeks, but made a permanent mark on the county's birding history!

Conclusion

When it comes to patch and local birding – be it for the first time or as an experienced birder – you will want to weigh up whether a single site or wider area is the most suitable way to go. Access and travel, plus birding potential, are important factors to consider.

Once you've worked out which road you want to go down, try and identify a site or area that offers a variety of habitats, or as many as possible. This will keep your birding dynamic, and will be useful in quieter times. It will also allow you to turn your hand to different forms of birding depending on the time of year and weather.

By using mapping software, or good old-fashioned paper maps, you can establish your patch or area and define its boundaries. You can also explore an area to get a feel for it and garner an understanding of whether or not it would make a good patch.

By this point, you have yourself a patch. You know what site you want to cover regularly or what your local area encompasses. It's time for the fun to begin – and for the birds to be found!

Patch tales: Inland seawatching

Matt Phelps is a birder based in West Sussex. The main focus of his many years in the field has been patch and local birding.

Like many birders, I keep lists. It's just something I've done for as long as I can remember, starting with the common species in the family garden, then adventures further afield. I've kept patch lists, work lists, garden lists, holiday lists. Perhaps rather strangely, the ones I have always felt the least invested in are my British list and various county lists. I do go on twitches occasionally – it's always nice to see new birds, enjoy the atmosphere, and catch up with familiar faces – but the most memorable moments have largely all happened close to home.

After bouts of childhood enthusiasm, I seriously took up birding in my mid-twenties, and my focus has primarily been local birding. Some of my fondest memories remain watching Swifts and House Martins over the family garden, and finding Firecrests and Red Kites in the nearby nature reserve. It seemed such an adventure to walk or cycle there, although it was barely a kilometre away.

Home for the past eight years has been the Arun Valley in West Sussex. I have Ramsar wetlands, chalk downland slopes and more on my doorstep. For the first four years, I visited RSPB Pulborough Brooks almost daily and found great birds including White-rumped Sandpiper, Iceland Gull and Common Scoter. Since 2022, I've broadened my efforts to a 10km radius from home.

This has transformed my birding – more relaxed, more confident, and ultimately more enjoyable. Scarce breeding species have always fascinated me, and in recent years I've discovered Honey Buzzard, Long–eared Owl, Lesser Spotted Woodpecker, and more breeding a short distance from my house. Then there were the migrants: Night Heron, Savi's Warbler, Green-winged Teal, Dotterel

River Arun

and Marsh Warbler, all just a few hundred metres from home.

Something about finding pelagic species on inland waters or overhead really thrills me. They remind us of Britain's small size and how seabirds frequently 'cut the corner' across land. I've been lucky over the years. Back in 2013, while working on a private estate near Guildford, I saw my first inland Gannet, drifting low over treetops one grey October morning – it remains etched in my mind.

From the tower of Leith Hill in Surrey, one can see Heathrow, Gatwick and the sea at Shoreham, and it is a favourite spot of mine for vis-mig. In May 2018, we picked up four Great Skuas ('bonxies') powering east before heading straight south – my only inland skuas to date, and truly unforgettable.

Above: For Matt, of the thrills of inland patch birding is the occasional encounters with coastal species, like Little Tern (top) and Sanderling (bottom).

Since 2021 I've regularly visited farmland in north-west West Sussex with a good-sized reservoir. It has yielded remarkable species: Spotted Redshank, Sanderling, Bar-tailed Godwit, Black Tern and even a wintering Long-tailed Duck – twice! In 2024, two species topped them all. One misty June morning, a Little Tern tagged behind a Herring Gull, circled the reservoir and flew north – my first ever at the site. Then in November, after Storm Bert, a lone slim-winged gull overhead caught my eye: a first-winter Kittiwake! Some 20km from the coast, it was my 171st and final addition to the year's local total.

Branching out from one site has helped me understand the landscape and its birdlife more deeply. That November Kittiwake flew over woods where I've watched Goshawks and Honey Buzzards, past fields where Dotterel and Long-eared Owls appeared. Local birding is a journey that never ends.

The excitement of potentially seeing something like a Long-tailed Skua or Little Auk at a familiar inland site keeps me going – sometimes in dreadful weather, often against the odds. But moments like those I've shared here are what make it all worthwhile. If you're thinking about getting into birding, my advice would be: start local. You never know what you might find.

Wryneck

CHAPTER THREE

Understanding time and weather

Wherever you patch, be it a coastal hotspot in south-east England or a quiet inland area in northern Scotland, it is possible to harness the weather and time of day and year to your advantage – and in patch birding every opportunity to improve your luck should be taken! This section will lay out what times of the day are optimum for particular aspects of patch and local birding, and explain how they can vary. As well as this, we will look at weather, and how certain conditions should lead you to visit particular places with certain birds in mind.

Weather can have a huge impact on birds. For example, mist and fog can disorient and ground migrants.

Changing weather patterns

It is important to caveat this section by acknowledging the ever-changing weather patterns experienced in the UK. The climate crisis means seasons are becoming less classically defined – with wet summers and mild winters becoming more regular, as are intense weather episodes. This impacts birds' behaviour and movements, and the reality of patch birding is that anything can occur at any time. Previous norms and trends are no longer to be taken as gospel, such is the unstable state of our planet's weather.

Time of day

It's widely accepted that mornings are the best time of day to get in the field. Birds are often most active around dawn, especially outside of the winter months, when temperatures are milder and the longer daylight means disturbance is reduced. During migration seasons, both nocturnal and diurnal migrants may be on the move in the mornings. Typically, the middle of the day sees the least activity – especially when it's hot. Dusk can be a productive time to see nocturnal species like owls. Ideally, patch visits will take

A seven-day tide forecast

Mon	Tue	Wed	Thu	Fri	Sat	Sun
Low 05:08 (2.41m)	Low 06:03 (2.58m)	High 01:09 (4.65m)	High 02:32 (4.85m)	High 03:37 (5.03m)	High 04:33 (5.32m)	High 05:22 (5.54m)
High 10:59 (5.25m)	High 11:56 (4.23m)	Low 07:09 (2.66m)	Low 08:16 (2.58m)	Low 09:33 (2.34m)	Low 10:41 (1.99m)	Low 11:40 (1.62m)
Low 17:43 (2.03m)	Low 18:46 (1.41m)	High 13:09 (4.98m)	Low 14:32 (5.13m)	High 15:36 (5.36m)	High 16:35 (5.62m)	High 17:29 (5.83m)
High 23:54 (4.79m)		Low 19:49 (1.98m)	High 21:07 (1.73m)	Low 22:15 (1.55m)	Low 23:17 (1.29m)	Low 00:11 (1.15m)

Knowing tide times is essential if you patch by the coast because they dictate where and when birds can access their rich feeding grounds and when they must roost to conserve energy.

place in the morning and, depending on the size of the area you're covering and how much time you have, will last at least an hour, ideally two or three. But various factors can influence one's ability to do this, not least the amount of free time you have!

The truth is, however, that even a snatched 15-minute visit in the middle of the day can produce good birding. So, try to visit your patch or patches as often as you can, regardless of the time of day or year. Any opportunity to get in the field is an opportunity to be taken and, indeed, in times of optimum weather (such as heavy showers in early May), it is worth seizing the moment, regardless of the time of day.

If you patch by the coast, then gaining an intricate understanding of tide times and heights is really useful, especially if you're watching an estuarine site. These will of course vary depending on where you live, so keeping track of tide times is important.

In the table below, the day is broken up into loose categories, with suggested focal areas for each. During the busier times of the year, multiple visits to your patch or local area are advised. Especially in the autumn, during busy periods of migration, it can feel like there is a constant turnover of birds. If you're looking for passerines, then an early morning and evening visit can be useful, either at the same site or across two different ones. If you cover a single site, then multiple visits are a little more straightforward; if you're working a local area, then try to visit your four or five key sites at least once a week during a busy migration period. Depending

Early morning	The best time of day to be on patch, year-round. During spring and summer, singing and breeding activity will be at its highest and, during migration, this is by far the optimum time, especially for vis-mig.
Mid- to late-morning	Generally quieter than early morning. The best time to look for raptors (either birds on migration or resident species) as well as other large soaring birds. During spring and autumn, it can be a good time of day to look for diurnal migrant passerines too, such as chats and wagtails.
Midday to mid-afternoon	Generally, the quietest time of the day for birding, but good for raptors and, during cold spells in the winter, sometimes the most active period for certain birds, especially passerines.
Late afternoon to evening	In the winter, a good time to look for gulls gathering at roost or pre-roost sites. During migration, many species will feed up ahead of a nocturnal flight and activity, especially among passerines, can be as high as in the mornings.
Dusk	Good year-round for owls and, in winter, for observing roosting flocks of various species.
After dark	During spring and autumn, a good time for sound-recording overflying migrants (nocturnal migration or 'noc-mig') – and if you fancy it, sitting outside and listening yourself.

on how much time you have, visiting multiple sites in a morning outing is recommended – providing you are able to cover each relatively thoroughly and are not simply racing between spots.

Time of year

The patch calendar section (Chapter 5) covers in detail what sort of birding tactics to employ at various times of the year. However, it must be stressed that certain months are worth investing more effort in than others. During spring and autumn migration, it's worth trying to get as much patch birding in as you can, be it a quick 15-minute check of a local lake or a four-hour vis-mig watch. The winter and summer can undoubtedly produce exciting birding, but they are also opportunities to take moments of patch rest and take your foot off the gas a bit. Summer in particular can have a birding 'off-season' feel to it. It is not worth overdoing it at such times.

Below are examples of days on patch (presuming you have free time) for each season:

Winter

- Morning check of a waterbody for waterbirds.
- Afternoon check of farmland, targeting finches and buntings for example, as well as hunting raptors and owls.

Spring

- Early morning check of a waterbody for general passage, including waterbirds and waders.
- Later morning check of a 'dry' site (for example hedgerows or fields) for passerine migration.
- Evening check of a waterbody for general passage and/or a 'dry' site for passerine migration.

Summer

- Morning visit to a species-rich site to log breeding records.

Autumn

- Early morning visit to a vis-mig site to watch the skies.
- Middle of the day visit to an open-country site for passerines and other landbirds.
- Evening visit to a waterbody.

Weather

Weather has a massive impact on birds and their day-to-day lives. It can influence feeding, nesting, roosting and flight behaviours. For the patcher, it can make a huge difference to the birds you see on any given day. Many birders keenly study weather charts and forecasts, as they try to predict what species might be found off the back of certain conditions. It's impossible for this section to cover every specific type of weather that can impact the birds in your area – there are too many variables for that, from both a weather and geography perspective. Many weather conditions that are broadly relevant are discussed in the calendar section – freezing weather in December and showery days in April, for example.

Weather is particularly pertinent when it comes to migration, which is a significant and exciting part of any patch birder's year. Migration peaks in the spring between April and May and in the autumn between September and October, but the reality is that movements of some form can

Wind strength will impact migrating birds. Hirundines, for example, usually fly lower in strong winds.

be witnessed year-round, even in the middle of the winter, particularly during cold conditions, and in high summer, when post-breeding dispersal commences.

Of course, elements other than weather are in play – and it's worth remembering that there are different types of migration, from long-distance to irruptions – but successful migration is centred on good weather. This, coupled with the bird's condition, dictate whether or not it will migrate at a given time. As a result, the variables are almost endless when it comes to migrant birds and, during the passage seasons, there can be significant differences in the volume of migrants from one day to the next.

Wind plays perhaps the most significant role (more on which below). By reacting appropriately to wind conditions (and taking advantage of favourable tailwinds), birds significantly boost their migration speed and reduce their energy usage. On top of this, they are less likely to be blown off-course.

Optimal conditions for the migrating bird probably make for a smooth, successful journey, but also mean that a patch birder is unlikely to detect it! Suboptimal migration weather is more likely to produce birds. Bad weather in the right place can mean that birds are stalled or even brought down to locations where they can be seen much more easily. As one example, Common Terns are often present on larger waterbodies inland, but at those same sites Arctic Terns are only likely to be recorded on migration when rain or wind forces them down low enough, particularly in spring. So, it is in fact bad weather for migration that should often be sought out. However, there are many intricacies, even on a local level. Therefore, noting weather and the birds you see off the back of different conditions is a fundamentally important way to build up your own personal understanding of your local area.

Wind

For a migrating bird, wind is the single most influential environmental factor – and so wind is an important element for a patch birder to consider! Birds prefer to travel with a tailwind and often gain altitude – frequently higher than humans can see – to take advantage of it. Indeed, various radar studies have shown that migrants can adjust their flight height to locate the level with the most favourable wind. Conversely, birds avoid flying into headwinds. For small species, even a light headwind can halve their range, while a strong gale can push them backwards. To survive, they fly low and keep close to the ground.

All of this can be difficult to interpret for the patch birder. Let's say it's the

Prolonged freezing conditions can send winter thrushes, like this Redwing, on long searches for food.

autumn. You want easterly winds to send birds over that are migrating south from Fennoscandia; you also want non-optimum conditions, maybe even a headwind, so they are low down enough to be detectable. Often a 'perfect storm' of conditions, both locally and further afield, is required to produce the best results.

Britain's position on the western edge of a continent is the main reason why easterlies are eagerly anticipated as such winds send birds over from Europe and beyond, or push migrating birds further west. Of course, fast-moving storms from North America can deposit Nearctic vagrants on our shores, from waders to passerines. But these are far from your everyday migrants. And different times of year can mean different things; northerlies in the winter can sometimes bring cold weather movements, southerlies in spring can produce migrants overshooting more southerly breeding grounds, and so on.

Many of the more exciting species come, in spring, from southern European and northern African climes and, in autumn, from Scandinavia east to Russia and Siberia. In spring, with birds racing north to occupy their breeding territories, long-reaching, warm southerlies will send the highest volumes of migrants on their way. If coupled with inclement weather in Britain, migrants will become grounded – and thus detectable. In the autumn, millions of passerines, waders and wildfowl head south across Europe, Central Asia and Siberia, and many make it as far west as Britain. Those from Scandinavia have generally arrived via eastward-moving areas of high and low pressure – and again, inclement conditions will ground birds upon their arrival. One famous autumn example came on 3/4 September 1965, when Suffolk received a 'fall' of extraordinary numbers. At Walberswick, this was said to involve 15,000 Redstarts, 8,000 Wheatears, 3,000 Garden Warblers, 1,500 Whinchats, 1,500 Tree Pipits and 1,000 Willow Warblers. On that occasion, clear skies over Scandinavia had created good conditions for migration, but a depression was moving up from Germany with rain and easterly winds on its northern edge, forcing masses of birds down on Britain's east coast.

Of course, these are only rules of thumb – and a huge range of wind directions can influence specific species. For example, spring north-easterlies might be good for Black Tern, Little Gull or Red-backed Shrike, while strong blasts from the south or north in late autumn might produce Pallid Swifts or Little Auks respectively. Similarly, migrating passerines and raptors are often more detectable in headwinds, despite this in theory being the least optimal weather for them to move. The best wind direction for vis-mig typically varies from location to location, too, and it pays to keep a track of the wind direction (and strength) on all of your birding sessions so you can study patterns of occurrence and identify the best conditions for your patch.

Cloud cover, precipitation and visibility

For many migrant species, the more cloud cover that is present, the more likely the birds are to be detected. Fine, sunny conditions with a light tailwind will probably result in many species migrating high overhead, unseen by us humans (although it's worth pointing

Wheatear is a classic sign of migration for many patch birders.

out that this weather is optimal for soaring birds, like raptors, storks and cranes, which require thermals to gain height). So, whether you're bush-bashing on the coast or checking an inland waterbody, cloud is likely to be your friend, especially during the peak passage times.

Cloud cover will force birds to fly lower than usual. In my area, the best autumn vis-mig conditions require some cloud cover – not only are the skies easier to scan, but the birds are flying much lower. In spring, cloud cover and, ideally, intermittent rain, is absolutely ideal (and will entice me swiftly to one of my local waterbodies, where waders, gulls or terns might be dropping in).

Rain is an interesting element of local birding. You want just the right amount. Persistent rain over long periods means birds simply won't move. Similarly, if it's raining in your area as well as wherever it is the birds might be arriving/taking off from, then there's unlikely to be much movement. Showery conditions are ideal – light or heavy – and normally are enough to drop birds down. Precipitation plays a lesser role in the autumn, with birds not in such a hurry to reach their end destination. As a result, rain is best avoided (though can still be productive for waterbodies).

Despite all of this, birds can be found in the most unexpected weather. I've had terns drop into my local reservoir on cold, bright and still mornings, and spent hours seeing nothing at similar sites on apparently perfect showery, grey days.

Another important element is visibility. Birds that have taken off in perfect conditions far away may suddenly hit fog or murk in Britain, and ditch down, sometimes in less than ideal places. In weather like this it is of course

The lunar cycle affects the nocturnal migration activity of various bird species.

not always easy to bird, but if visibility is OK, then it's worth getting in the field. Such weather is particularly good for waterbodies, perhaps dropping down wildfowl like Common Scoter or passage waders. It's also good for passerine migrants, especially in coastal areas.

Temperature

Temperature plays a role in bird migration as well. Freezing cold weather in December, be it on the near continent or within Britain, is likely to set wildfowl, waders and thrushes on the move. In high summer, heatwaves can shrink wetland habitat and have a similar impact, turning certain species into relative nomads.

During the passage seasons, the temperatures are generally balanced, but there are subtle nuances that can be appreciated. For example, a muggy, mild spring morning is likely to be better for passage terns and waders then a clear, cold one; on the other hand, a cool, clear day with a light headwind might be better for migrating raptors.

Interpreting weather maps

Understanding how to interpret weather maps can be incredibly helpful when predicting the best conditions for patch birding. These maps show meteorological features across a particular area at a particular point in time. The lines on a weather map resemble the contour lines of an Ordnance Survey map. However, on weather maps, those lines are isobars, which connect locations with the same average air pressure (which is measured at sea level). The closer the isobars are to one another, the stronger the winds. Flowing arrows indicate the general direction of air currents, although local variations may occur.

Isobars enclose areas of high and low pressure, labelled as 'H' and 'L' on the map opposite. High-pressure areas, known as anti-cyclones, are characterised by light winds that can strengthen or weaken at any time. The edges of these high-pressure systems often bring strong winds, as they compress the isobars together. The centre of a high-pressure system may also contain low cloud, leading to frost in the winter and thunderstorms in the summer. Very high-pressure areas can block the progress of weather fronts behind them. Low-pressure areas, known as depressions, are typically associated with cloud cover, though these clouds are generally higher up. Air masses come from various parts of the world – they may be warm (from the tropics), cold (from the Arctic), moist (from maritime regions) or dry (from continental areas).

A front is the boundary between two different air masses. A cold front, which represents the leading edge of a cold air mass, pushes warmer air upwards, creating clouds and rainfall. It is shown by a line with triangles pointing in the direction it is moving. A warm front, the leading edge of a warm air mass, rises over colder air, often producing clouds and heavy rain, as well as shifts in wind direction. It is represented by a line with semicircles pointing in the direction of movement.

When a cold front overtakes a warm front, trapping the warm air above and causing it to cool, thick clouds and rain form, resulting in an occluded front. This is depicted by a line with both semicircles and triangles pointing in the front's direction. As the front continues to move, temperatures stabilise, rain becomes more scattered and winds calm. A stationary front, indicated by triangles on one side and semicircles

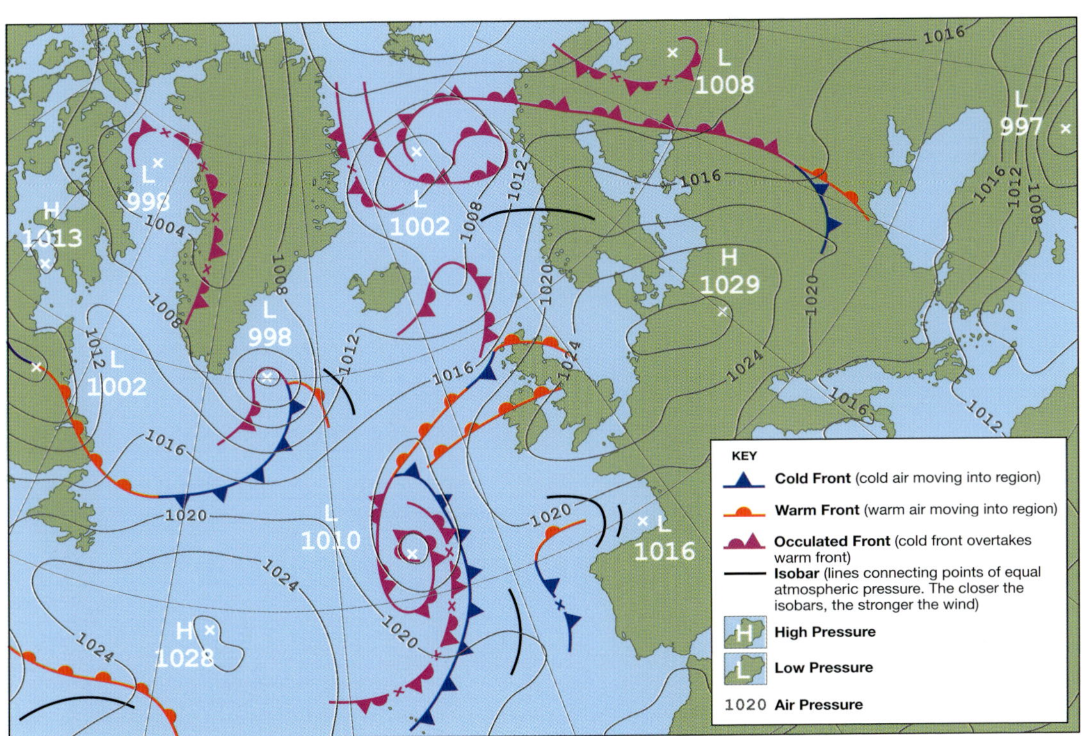

Being able to read and interpret weather charts can help you predict when periods of migration may occur.

on the other, occurs when a front stalls. Rain can persist in such conditions until the front moves along.

Getting your head around weather maps and charts is particularly useful during periods of migration and may lead to you maximising your time on patch, allowing you to tie in your visits with the most exciting-looking conditions.

Conclusion

Ultimately there are huge variables when it comes to patch birding and the weather. There are good rules of thumb to practise and methods to deploy, but birds will often defy apparent logic. Ultimately, getting in the field whenever you can is the best tactic. But it is worthwhile to try and understand what conditions drive certain species into your area. Elements like wind direction, cloud cover and rain will vary from one site to the next in terms of how they relate to birds – it's impossible for this section to provide an exact answer. However, by keeping track of the weather on each of your patch visits, watching the forecasts and studying the charts, you will be able to learn how to target good-looking conditions for certain species at certain times of the year.

Above: Meadow Pipits are one of many species that show a clear spike in migratory activity in spring and autumn. Large movements are often dictated by the weather. **Left:** Low cloud and rain has a bearing on the migrating birds – a heavy spring shower might drop in a wader or two to your local waterbody, for example.

Patch tales: Ringing the changes

Dan Owen is a keen local birder and ringer based in Cheshire, with Woolston Eyes his cherished patch.

I first visited my patch – Woolston Eyes Nature Reserve, Cheshire – in the winter of 2015, to watch an enormous Starling roost that had formed. Estimates of up to 400,000 Starlings were made that winter, and from then on, I was hooked. Indeed, 10 years later I still spend more days there than not!

Patch birding is about building a deeper relationship with a single location, something that resonates with me more than globetrotting to tick as many species as possible. This local approach encourages familiarity and a sense of responsibility for the site, and is altogether more fun to me. The element of surprise is one of the most rewarding aspects of patch birding: watching new behaviour of a common species or stumbling upon a rare vagrant both feel more significant when you've put in the extra effort. I'm also fortunate that patching Woolston comes with a sense of community, joining others who have been watching the patch for nearly 50 years! Contributing to science and to understanding of the place is really important to me.

The sense of familiarity with patch birding has great appeal; for example I know which pools to check when the first Garganey are due back, or where to look for passage Spotted Flycatchers. However, each year is different and a White-tailed Lapwing calling in one day late in May wasn't something any of us were expecting – likewise the arrival of several Ferruginous Duck in recent springs!

Dan's Woolston Eyes patch is a mix of habitats, situated on the edge of a large urban area.

Late spring and early summer is one of my favourite times to be on patch, for the breeding season. I'm fortunate to have many scarce and rare breeding species on patch, and it's always a joy to confirm their presence for another season. Watching the Black-necked Grebe colony is always a privilege, and each year is different from the previous. Another fond memory was proving breeding Garganey, the first time this had been confirmed in the county in 50 years! After much effort and perseverance, I eventually located the young – such a satisfying feeling.

Autumn comes with its own surprises too. A Little Bunting trapped and ringed on site was a real shock, and finding the county's second and third Penduline Tits just a couple of years apart was very exciting. One of my fondest memories is a large influx of Yellow-browed Warblers onto the patch in autumn 2024, when I found no fewer than 16! Despite having seen triple figures of the species nationally that year, including in places such as Shetland, I know which I preferred. A flock of 10 Greenland White-fronted Geese was also a nice treat during an autumn vis-mig – the first double-figure flock in England for a good few years. Thankfully they flew over just as I was packing up for the morning – talk about luck.

Watching the changing seasons through patch birding gives a great sense of connection to the rhythm of the site; watching migration, breeding patterns, and any shifts in behaviour and populations give a sense of accomplishment. The sense of anticipation with each turn in season, and tracking any such changes, are very fulfilling and enrich both my personal enjoyment and ecological awareness.

Dan has good form for finding rare and unexpected birds on his patch, including multiple Ferruginous Ducks (left) and two Penduline Tits (right).

Wheatear

CHAPTER FOUR

How to get the most out of your patch

You've identified and defined your new patch or area – and now you are ready to make the most of it. You've already brushed up on the optimum times of day and weather conditions – so what else can you do to push your patch efforts to the next level? In this section we'll discuss the various ways you can maximise your efforts and get the best out of your patch birding.

Challenge yourself

As soon as listing is mentioned in birding terms, some birders switch off. But it's not all about competing with others or racing around after ticks. As with any avenue of life, setting yourself targets is a really good way to push yourself – and it's the same with local birding. Setting a certain year list target or simply keeping track of how many birds you've seen is a great way to keep yourself focused and hungry.

Whether you're already familiar with your patch/local area or not, you can work out roughly how many species could or should be expected in a certain year. And, based on that, you can set a bar to reach. Depending on how much effort you plan on putting into your patch efforts, you could write down every species based on different categories (guaranteed, likely, possible, unlikely, very unlikely) and whittle through them as the year goes on.

Golden rules

Here are what I consider to be five golden rules of patch birding:

Time = results Patience is arguably one of the most important traits for any birder. Patch birding patience is perhaps a little different, though. Anyone planning to go looking for birds at a site or sites close to home (especially inland!) must accept that there may be many hours of fruitless scanning, seeing the same species time and again before 'the big one' arrives. In patch birding, however, with patience comes rewards.

Anything can turn up at anytime, anywhere Even during the quieter times of the year, or on less productive parts of your patch, a good bird can appear. Every day is different. As a result it's important to look at your patch with fresh eyes – expect the unexpected always; never be complacent, as often the most surprising bird will be found at the least likely moment.

If an approach delivers, it's worth repeating Whether it's a new spot that's proven productive, certain weather or a specific time of day that's delivered, or a fresh style of birding that's yielded results, try it again!

Learn which weather conditions work, and which birds to look for A clear understanding of weather patterns and how they might influence bird movements is crucial to being a successful patch birder. Many migratory species are heavily affected by weather, and learning to read the weather charts can help determine where might be best for you to head out on a certain day. At the same time, keeping abreast of which species are moving through nationally and regionally will help you to better target particular areas within your patch.

Stay optimistic There will be periods where you feel you are finding nothing, the good birds are turning up elsewhere and you may even question the point of what you are doing! But think back on all the good records you have had previously – rest assured that there will always be more to come. The low points are just that and it's important to not get stuck in a rut.

Building a patch year list can be a fun way of providing some structure to your birding efforts. Whooper Swan is a relatively easy to find winter visitor in most parts of the British Isles.

It's a good way to keep you motivated – and by seeking out certain species and getting in the field, you are always likely to bump into something else, be it an unexpected bird, a high count of a common species or perhaps a pleasing breeding record.

In future years, you can try and beat your personal records. You'll find yourself grabbing small 10- or 15-minute windows of opportunity to try and find that target bird, not knowing what such a flying session could produce. My personal record year list for my south-west Surrey local area is 163. A decent effort for a landlocked region, but I know other birders who have achieved far higher tallies in inland regions. It just goes to show that, with a little effort and desire, a single year in a local area can be productive – and full of different birds!

You can co-opt big days into such efforts too. Spring is often a popular time for bird races (when birders try to see as many species as possible in a single day), and the preparatory work and understanding of expected species are fun parts of the process (see more on page 101).

In a similar vein, breaking things down to even more granular levels, such as month listing, can be interesting and insightful. Certain apps, including eBird (see page 172), record how many species you've seen per month in a certain area. I personally find it fascinating to compare the different levels of diversity each month offers, and it provides data on what's unusual at certain times of the year and common at others.

Understand and appreciate rarity context

Understanding what's unusual in your area, be it generally or at different times of the year, is a hugely important element of patch birding. If you go into it looking at rarity value through a national lens, then you're likely to feel unrewarded for long spells – the reality is, national rarities will grace our patches very infrequently indeed. They're rare for a reason!

However, certain species that may not initially seem rare, or that may not be rare in a wider regional or national context, may be incredibly unusual on your patch or local area. By appreciating this, you'll be able to enjoy more regular thrills when encountering unexpected species. One random example would perhaps be Gannet – a bird that is rather common offshore almost anywhere in Britain. However, if found on your local inland reservoir, it would represent a very unusual occurrence. Contextual rarities like this are almost innumerable. Such birds can even be broken down somewhat into categories, as in the table below.

Being able to grasp what is rare or not does require an understanding of your local area. So, immerse yourself in the information that's available – 'learning local'. Most counties have bird clubs that produce annual reports or even avifaunas, and these can be absolute treasure-troves when it comes to educating yourself on the status of birds in your local area, from rarity levels to previous high counts, as well as what breeds locally and what doesn't. Furthermore, most counties have a list of county rarities that require descriptions – these can be worth familiarising yourself with to understand local levels of rarity. For example, in

Local rarity	A species that is rare in your area, regardless of its wider national status. So, a Gannet on an inland waterbody, or a Dartford Warbler in some scrub in northern England, would fit this. In short, a bird that might be common in parts of Britain, even in places not far from you, but one that's rare in your specific patch or local area.
Patch rare/ patch gold	This is more specific to single sites and generally involves a bird that is common in your local area, but is particularly rare at a certain site. It might be a Coal Tit in a small block of woodland or a Great Crested Grebe on a pond you visit often. Generally, you'll need a good volume of previous visits (or extensive historical data for a site) to understand how truly notable a patch rare/patch gold is.
Seasonal rarity	A species that may be common at certain times of the year but is rare at others. An example in my area is Common Gull – it is one of the numerous gulls from October to March across my region, but from April–September there are barely any records, so a midsummer occurrence is highly notable.
Breeding rarity	Confirming breeding on some level of a species that rarely or doesn't usually breed in your local patch, area or region. One example in my local area is Teal – a common winter visitor to Surrey but rare in the summer, with only one or two sites usually producing breeding records in a given year.
Big counts	High counts of certain species, common or not, require meticulous note-keeping over longer periods of time, but are rewarding when stumbling upon a significant gathering of birds. For example, you might 500 Stock Doves on some farmland and, looking back through your records, see that such a figure is unusual for your local area.

Surrey, Merlin, Whooper Swan and Rock Pipit are county rarities – species that are common or at least far more expected in other parts of Britain, but noteworthy here. County rarities also represent satisfying finds!

Public bird-record databases, such as eBird, are also valuable tools when it comes to understanding what's rare and what's not. These are by no means complete, and sometimes inaccurate reports can slip through the net, but they're still useful.

And, of course, meeting and chatting with other local birders is a great way to improve your understanding of the birds in your area.

Plug in

Being able to maximise your time on patch means learning what's going on in the birding world on a wider regional or national level. Identifying promising weather patterns and how to exploit certain conditions for finding birds is covered in a previous chapter, but there are many other ways of plugging into the 'bigger picture'.

For starters, it can pay to engage with local and county birding scenes. This may be possible on social media, perhaps Facebook, Bluesky or X, and many local birders have WhatsApp or Band groups for instant messaging of bird news. This helps paint a picture of what other people are seeing – and what might be worth looking for in your area. For example, on a day of heavy Redwing passage in October, one can be prompted to take a look outside on the back of what other people are reporting.

Being tuned in to news like this is very useful, and subscribing to national bird news services can be worth it on an even broader level. If, say, a fall of Pied Flycatchers is underway one August day, this can be your cue to go looking in your local area for one as soon as you can.

The volume of bird news available today – and the range of sources from which it comes – can be a little overwhelming. It can even be dispiriting at times, especially if it feels like everyone

Being tuned in to the latest bird news in your area means you can make the most of other people's finds. Here birders watch a Short-toed Lark in Surrey.

bar you is seeing something interesting. But it's worth connecting with the local ebbs and flows of bird movements, as it can prompt you to get in the field.

Submit your records

Accurate bird records are the backbone of ornithological study. Documenting the birds you see is beneficial and fascinating from a personal point of view, too. Sharing those records with organisations such as the British Trust for Ornithology (BTO) or your county bird club is important and contributes to a greater understanding of population trends of certain species – and may even influence conservation action.

Tucked away in my loft is my first ever bird notebook, in which I began detailing the birds I saw from the age of eight. Ever since then, I have kept a log of my sightings – admittedly in varying levels of detail and effort. I have always enjoyed note-keeping and love that notes allow me to mentally time-travel back to different moments, places and birds.

The changing status of bird populations is one reason why submitting your records is important. Red Kite is a species that has increased greatly in the British Isles.

Naturally, my records have evolved into digital ones. I now have my thousands of checklists from different birding sessions, spread over the course of two-and-a-half decades, all online and within easy reach. A big birding change in recent years has been the conversion from notebooks to digital archives – and I know birders who have struggled to part ways with their beloved pen and paper methods. But in the modern world, doing things digitally is the most logical way to go, for you the birder and for the people retrieving and using the data at the other end.

Even in my own notes I can see the changes that have occurred in my local area during that time. The excitement of seeing my first local Red Kite, for example, and the greater regularity with which I heard singing Cuckoos in the late 1990s and early 2000s compared with the modern day. These sightings reflect the changes within bird populations all around us and, when you take all of this data from all birders in a certain area, region or country, form a hugely powerful store of information.

Recording what you see and hear is a hugely important part of patch birding. I have met plenty of birders who either don't want to record what they see or don't want to share their sightings with anyone else – and there is no doubt that simply connecting with nature is fantastic (and should always be guilt-free). However, to my mind, if you're a relatively active birder and are in the field fairly often, your sightings can play a truly valuable role in painting wider local, regional and national pictures.

Furthermore, I believe thorough record-keeping refines your skills as a birder. The birders I know who use BirdTrack, eBird or their own record-keeping system (see page 58) tend to pay far greater attention to detail and are more likely to look for things like breeding behaviours, unusual counts and early or late dates than birders who don't. Certainly, since I moved from notebook to eBird several years ago, my local birding efforts have increased – I'm sure I am encouraged outdoors more and scrutinise even common species with greater energy than previously.

But what is the best way to contribute to the 'greater good' of citizen science? If you are going to go to the effort of submitting records, you'd like to know that they are going to end up somewhere useful. There are various ways to keep and share records and we will take a look at some of them below. The first port of call is your local bird club. From there, any rare breeders or scarce migrants will be filtered and go on to the Rare Breeding Birds Panel (RBBP) or published in the annual Report on scarce migrant birds in Britain in *British Birds*. Other sightings may be used in publications such as county bird reports. These days, it is easy to log your records into apps and websites that simultaneously send them to your county bird club and the BTO. Given that this is the ultimate goal, their usage comes strongly encouraged.

And, of course, you may want your records for yourself. If nothing else, they serve to enhance your trip down memory lane when you remember good birding days in years to come.

Be a good note-taker

Where possible, it's important to try and add as much detail as you can to unusual or notable observations. When it comes to breeding records, most platforms have different codes to represent what level of breeding you have observed, from a singing male through to an adult feeding a chick. It's helpful to use these, even for common species. Extra comments on things like early or late migrants, vis-mig movements and high counts are encouraged too. It's certainly fascinating (and sometimes worrying) to compare your personal early and late dates of spring migrants, for example, and see how some birds are shifting their habits in line with climate change.

Counting birds is a really good practice to get into as well, not only from a personal satisfaction perspective, in terms of noting high counts, but also

Careful note-taking can help document rare breeding birds, such as Spoonbill.

to contribute more valuable data to the broader picture. Even if you aren't going to count every Blue Tit or Magpie you see, try to roughly estimate how many you've seen – a much more useful entry than leaving the count field blank or simply putting 'X'. You can find some guidance on how to count birds at eBird: support.ebird.org/en/support/solutions/articles/48000838845-how-to-count-birds. Logging the birds that you hear calling and singing is also valuable – I like to take a few 'breaks' while on patch in which I stand silently and listen to the soundscape, logging the common species I can hear. Steadily recording regular birds can be rewarding too. Many rarities have been found by those who put in the time counting through the common stuff.

Finally, it's worth mentioning weather as much as you can in your notes. This will help build up reference material for certain conditions and patterns that can produce migration or unusual birds (see Weather, pages 39–47).

Different platforms

BirdTrack
Run by the BTO, BirdTrack provides facilities for birders to store and manage their own personal records as well as using these to support species conservation at local, regional, national and international scales. Important results produced by BirdTrack include mapping migration timings and monitoring scarce birds.

eBird
eBird is run by the Cornell Lab of Ornithology in Ithaca, New York. Hugely popular in the USA, it has gained significant usage in the UK in recent years and, simply, offers greater functionality to the user as an app and a broader data portal than BirdTrack. Notably, you can archive photos and sound recordings. There is also a lightly competitive element to eBird. It tracks how many days in a row you've submitted checklists (I know people who have completed runs of thousands of days!) and how many species you've seen in a certain region or country in any given month or year.

County bird club
Submitting your records directly to your county bird club or county recorder is also an option. You might present them as a spreadsheet, or even provide copies of handwritten notes if you prefer. Note that BirdTrack records go straight to the county bird club/recorder – and that eBird records are publicly accessible.

Other options
Other apps and websites include iNaturalist/iRecord (which I use for my mammal and butterfly records) and Trektellen (a website focused on vis-mig).

County and local bird reports are a good way to build knowledge of the species in your area.

Reach out

Contributing isn't only about keeping and sharing your records. Try and engage with landowners – if you bump into a farmer or gamekeeper while in the field, have a chat with them, and show them your passion and knowledge. This can open doors, in terms of things like access to private land, but can also raise awareness of what birds might be on people's land that they don't know about. Generally, I find, such people have at least some interest in birdlife, and engagement with them on my patches has resulted in all sorts of perks, from specific habitat management for certain species to the pausing of particular farm work for others, as well as access to private land for things like vis-mig and ringing.

If you have a local or county bird club, and have the time, volunteer to help out. All organisations such as these are run off the back of selfless voluntary work and clubs are always looking for an extra pair of hands. If a local group is running field trips, consider going on one to meet other local birders – or even leading one yourself. And if there is a small group of birders in your area, why not set up a casual group, perhaps on a social media form such as WhatsApp or Band? Encouraging others and inspiring passion for a certain area is a positive thing. Being connected to a local site or area doesn't only have to be about studying birds.

Archiving

If you're particularly keen – and have time – it can be fun to sift through the history books to really build up a complete picture of your patch or area. Many patch birders like to look through old county bird reports and other local publications to put together a complete historic site list. It can be fascinating to see how the status of certain species has changed in your area – and there will probably be some surprising rarities in the mix as well! Creating such lists helps to reinforce your understanding of what's rare and what's not, as well as opening your eyes to possibilities that you may otherwise not have considered.

Create a micro-patch

Finally, I revert to actual field birding. A great way to squeeze every last drop out of your local area is to create a micro-patch. An increasingly popular concept among patch birder is the '1km' – a very small area of roughly 1km radius from one's front door. It's not your only birding area, but more the birding definition of 'on your doorstep'. You may be amazed at what does turn up so close to home when you take a closer look.

If you have a garden, then your 1km includes that, so birds that visit your feeders or fly over are included. And take a look elsewhere in your radius. Even

Unassuming areas on your doorstep can sometimes surprise you in terms of the birds they produce. For example, this small hill has proved a fine vis-mig spot.

if it's built up, there will doubtless be some islands of green – perhaps a small park or pond, or a bit of raised ground or stretch of river. These areas won't become your priority patch sites or take up much of your birding time, but if you're able to take a short trip outdoors, perhaps walking the dog or going to the postbox, that is a visit to your 1km. This will reinforce your understanding of the birds in your wider area while also allowing you to patch on your doorstep – and all without getting in a car.

Setting yourself a challenge, such as trying to see a certain number of species in your micro-patch, is good fun and you may be surprised by your results. The 1km is something I've adopted in recent years and really enjoy. I'm lucky to have a river running several hundred metres from my garden, and an area of open grassland and meadows less than two minutes' walk from my front door, but I've still been surprised at how productive the birding has been right here close to home, from cool breeding records to pulsating mornings of migration.

You may be lucky enough to have a large garden, too, in which case you might want to concentrate your micro-patching efforts there – it certainly feels that bit more special when a bird nests or visits your own little plot of land. Depending on how big your garden is and how dynamic the habitat is (and how close you are to productive bird areas), you could spend hours watching the birds in your backyard. Noc-mig adds an extra dimension, too, with night-migrating birds like wildfowl and waders possible additions to your garden list. Nobody else is going to record the birds in your garden, so it's an area that would otherwise be data-free when it comes to bird records. The ease of opening your back door or looking outside the window, mug of coffee in hand, is very appealing – and on several occasions has taken precedence over visiting one of my patches!

In short, local birding shouldn't stop when you leave your patch and head home – you should always be plugged in and, by paying a little bit more attention to your immediate surroundings, you may be rewarded with some pleasant surprises.

Sometimes unusual birds can turn up in the strangest places, like this Oystercatcher on the grassy banks of a small fishing pond one May morning.

Patch tales: Playing the weather

Josephine Snell is based in Surrey. Relatively new to birding, she has found patch and local birding to be the most fulfilling and enjoyable form of the hobby – and is now hooked.

I've only been birding for eight years, so I still consider myself a relative novice. Early on, I learned the basics at nature reserves and dabbled in bird photography. A neighbour introduced me to twitching, and I started county and UK lists. But I soon found myself wanting more: a deeper knowledge of birds, and a more enriching, environmentally conscious way to bird. That led me to patch birding in Surrey, where I could combine local exploration with chasing county lifers.

Instead of focusing on a single patch, I chose several under-birded ones with varied habitats. This approach rapidly improved my knowledge of species distribution across seasons and helped me understand how my patches fit into the broader Surrey birding context.

One early highlight was on a cold, clear morning in November 2021 at Thorpe Park. The pits surrounding the theme park are great for wildfowl but seldom birded. I'd already found a female Scaup that winter, and on this day, while scanning flocks of Tufted Duck and Pochard, a drake Ferruginous Duck appeared in my scope, glowing like burnished copper in the sun.

This wasn't just a county rarity but a national one. I put the news out on social media, and the duck was soon relocated by others. It overwintered at the site and was later accepted by the British Birds Rarities Committee (BBRC). This discovery reinforced the simple joys of patching under-birded areas and

Josephine found a Ferruginous Duck at a little-birded wetland site, showing that great finds can occur anywhere with a little effort.

methodically scanning large flocks – you never know what's hiding in plain sight. And perhaps most encouragingly, you don't need decades of experience to make an important find.

The following spring I began to seriously attempt spring-migration birding, but I was feeling disheartened – struggling with how to balance patching with chasing county lifers, and not finding much compared to others. A mentor's advice to "play the weather" stuck with me, though, so I visited a small reservoir – one of my patches – on a promising morning for waterbirds.

At first, nothing of note appeared, but after 90 minutes, a Sandwich Tern suddenly landed on a buoy in front of me. In most places, this wouldn't be extraordinary, but in landlocked Surrey, they're noteworthy – and it was a county lifer for me. I quickly got some photos, alerted the local birding group, and then it was gone, present for no more than 20 minutes.

This moment showed me I was progressing as a birder: I'd interpreted the conditions correctly and was rewarded with a rare bird for the county. It was also a lesson in perseverance – lows are inevitable in birding, but the highs, when they come, are worth it.

I wanted to include some good arable farmland in my patch network – hard to find locally, but after scouring maps I found a promising spot and began regular visits. A couple of springs ago, I was amazed to discover Lapwings displaying in one of the fields. They are in national decline and scarce breeders in Surrey. One female began nesting near a public footpath, and when I told the farmer – who was interested in wildlife – he took immediate protective measures: signage for dog walkers and stakes to keep farm vehicles away.

Studying rare and declining breeding birds, like Lapwings, is one of the main patch birding joys for Josephine.

Sadly, that nest failed. The field had dried out, and Lapwing chicks need soft, damp ground to forage. The pair nested again in a nearby field and hatched two chicks – but again, these didn't survive. I spoke with the farmer, explaining the habitat requirements, and he dug a shallow scrape to retain moisture for livestock – and hopefully the birds.

The following year, the Lapwings returned to the same field. This time, their chicks hatched and, guided by their parents, foraged in the newly dampened area. Over the weeks, I watched them grow and eventually fledge – one of the most fulfilling experiences of my birding life.

Patch birding can enable meaningful conservation action at the hyper-local level. By building a relationship with the land and its stewards, even casual birders can make a tangible difference.

Barn Swallow

CHAPTER FIVE

A patch birding calendar

Every month of the year can be a highlight for the patch birder. From crisp winter days scanning through wildfowl and those early spring mornings hoping to see your first spring migrants, through to lazy days of summer breeding activity and frantic autumn migration sessions, the local birding calendar is something to cherish. Every year brings its own wonderful moments, unexpected surprises and variety of birds. At any time, in any month, that one big moment could be just around the corner.

01 January

January is always a significant month in the diary of a birder. A New Year marks a new chapter and, even if you don't keep a year list, there is a feeling of resetting and refreshing ahead of the upcoming 12 months of birding. It's a time when plans are hatched, efforts are upped and optimism is generally bubbling away nicely.

The reality, however, is that midwinter in the British Isles is not exactly a peak time for species diversity and movement. With the days short and the weather rarely nice, it pays to play smart in January, seeking out trickier resident species, maximising winter habitats and plotting your year ahead. That said, there is always plenty to see during the depths of winter – particularly if you're blessed with wetland habitat – and January marks a great starting point for any patch birder, whether you are new to the game or are getting ready for yet another year of local birding thrills and spills.

New year, new lists

For many patch birders, January marks the beginning of a new year list. Setting yourself a year list goal is definitely something to consider ahead of January getting underway. It's not all about numbers and chasing birds around – year listing locally often gives you much-needed motivation to head out and look for species that you perhaps otherwise wouldn't seek out.

It's also fun to work out how much you can get out of your local patch or area – you might be surprised at the total number of species you can record in a single year. By mapping out your target species, perhaps on your computer or in a notebook, and chalking them off during the next 12 months (and not to mention the unexpected bonus birds you'll

Cold weather in January is a good time to search through flocks of finches and buntings, such as these Yellowhammers.

encounter), you'll find yourself working within a satisfying patch structure. Every year is different, which is one of the beauties of local birding – there will be bonus birds, as well as species you miss that you expected to see. Depending on where you patch, there may even be a history of year listing, and that can mean records to try to beat!

Of course, if you're year listing, then everything is new on 1 January. The short period at the start of the year, which places your everyday species on the same pedestal as much rarer birds, is always fun, and there may even be some common birds that take you a bit of time to pin down. This feeling of resetting helps contribute to an energised start to a new year and is one of the reasons why keeping a year list is worthwhile for the patch birder.

There are lists within year lists, too. Maybe you'll want to try a garden year list in order to spend more time close to home, or a green list that only features birds you've seen on foot or bike from your front door. If finding rarities is your thing, then a self-found year list is another thing to consider. Whatever your chosen year list might be, it'll give you extra impetus to get outdoors during what can often be a rather downbeat winter month.

Eye for extreme weather

In the modern day, winter weather in Britain is variable, to say the least, and can flip-flop from one extreme to another in the space of a few days. Certainly, prolonged cold periods with ice and snow are becoming less regular, at least in southern and western England. Meanwhile, mild, wet and sometimes stormy weather is becoming more expected – and often several stormy spells come within a short, intense period. These extremes will only become more pronounced as the climate crisis worsens in the years and decades to come.

Sewage farms are excellent places for insectivores to forage during the colder winter months. On a mild, still day, they can be a hive of activity and may even yield scarcities.

As a patch watcher, life's about making the best of what you've got, and that means using the weather to your advantage. In January, there are perhaps two types of weather to bear in mind. Hard weather (as covered in more detail on page 155) during the midwinter period can liven up local birding. Waterbodies freeze over, pushing waterbirds to congregate at remaining unfrozen sanctuaries, and in particularly hard conditions winter vis-mig efforts may see flocks of thrushes, Skylarks, Lapwings and Golden Plovers rove around, while winter gatherings of seed-eaters will increase in areas where food can be found (such as winter crops and gardens). Typically, the second and third days of prolonged cold spells see more significant movements of birds, though even a couple of days of cold weather can shuffle the local pack.

Another increasingly significant winter weather type to consider is heavy rain, which can lead to flooding. With winter rainfall levels on the up in Britain, places that would normally remain water-free are becoming more susceptible to floods. Areas of low-lying farmland, as well as river valleys and meadows, are worth checking during times of heavy rain. A rather empty, lifeless field may, in the space of a few days, be transformed into a wetland haven holding wildfowl, waders and gulls. In my area there is a particular stretch of water meadows that is, for much of the year, a quiet bit of grazed river valley. However, after heavy rain the area becomes submerged and, out of nowhere, the birds move in, a sudden congregation to sift through and enjoy.

January is as good a time as any to check such floods for birds like Cattle and Great White Egrets, and Glossy Ibis – especially in mild winters. From the other end of the compass, Whooper Swan and Pink-footed Goose – species that are enjoying breeding population booms in Iceland – may spread out into areas they haven't historically wintered in. So, keeping an eye out for sudden flooded fields, especially in areas of extensive farmland, is good practice. Such transformative conditions can completely change the mindset of the local birder when they're presented with a 'new' site. A good example can be read in Josh Jones' account of his year on patch (see page 182).

Winter warblers

On pleasant, still days, January is an ideal time to look for wintering warblers. With increasingly mild winters, especially in England, these birds are ever more regular – and scarcer species are increasingly seen. It is relatively straightforward to identify their preferred habitats, making them a tangible target for practically any patch birder. Two resident species – Cetti's and Dartford Warblers – are present within their ranges year-round, and Blackcaps are winter garden staples for some of us. However, it is wintering Chiffchaffs that you'll want to seek out because, if there's a scarcer species or subspecies to be found, it'll likely be among them, or at least in similar areas.

Areas around water are without doubt the best places to start looking, with one type of site particularly fruitful: sewage works. These are great places to look for warblers, with a ready supply of insect prey during the colder months. At this time of year, the temperature of the treated water from the processing works is a few degrees higher than the surrounding environment and, as a result, provides a microclimate favourable to insects and other invertebrates, which act as a vital food source for warblers.

Many sewage works can be difficult to access, but often you only need to be able to peer in to a site, or even bird the periphery of it. Surrounding vegetation and trees are the best areas to search, although often birds will use the actual filter beds. It's worth perusing your local OS map to find sewage works you may not know exist – you might be surprised to see how many there are in the British countryside!

Other sheltered stretches of water may also hold wintering Common Chiffchaffs, from trees and bushes fringing a lake or reservoir to shallow ditches and streams running through woodland – anywhere that supports insects at this time of year. Such places can even be found in urban areas, like parks.

The most likely prize among any wintering Common Chiffchaffs is the subspecies *tristis*, informally known as 'Siberian Chiffchaff', This striking form makes a satisfying midwinter prize (they're a personal favourite of mine). Yellow-browed Warblers are wintering in ever-increasing numbers in Britain and are also realistic prospects, especially close to areas of water and in southern

'Siberian Chiffchaff' is a prize find in the winter months, with sheltered spots with a rich insect source – such as sewage works – the best places to look.

and western parts of England (see more about this species on page 143). Even bigger prizes are possible, though of course are not to be expected – think Dusky, Hume's and Pallas's Warblers.

As mentioned, mild days are best. The obvious logic is that more insects will be on the wing. Bright days are also good, especially at the back end of winter, as some birds may be tempted into song, but still, warm and somewhat overcast conditions are likely to be the best mix – clear winter days can often be cold so it takes longer for insects to become active.

Garden goals

For many, midwinter isn't the time to frequent your garden (if you have one). It is worth keeping an eye on your outdoor space at this time of year, though, and January is a good time to get in your garden – the ultimate local birding hotspot! As harsher times arrive, more birds will utilise gardens for an easy meal. If you feed your garden birds, then January is an important time to regularly clean and top up your feeders. Try planting native species like ivy, hawthorn and holly. Their fruit provides a natural source of food for birds in winter.

Extremely rare birds have appeared in gardens before. This Dark-eyed Junco was in a London garden in 2021.

Occasionally your garden may be visited by something unusual, like a Hawfinch.

Also, you might just entice an unusual visitor that could represent a significant local birding prize. Gardens may not feel especially central to the patch birding experience, but the list of national rarities that have been found in British gardens is extraordinary, and includes Ovenbird, Dark-eyed Junco, Baltimore Oriole, Northern Mockingbird and Dusky Thrush, to name a few.

My garden is modest, and I've not lived in this house long at the time of writing, but I've had several surprises both in and over it, including Great Crested Grebe, Marsh Harrier, Whimbrel and Hawfinch! As suggested previously, why not keep a garden year list? If nothing else it'll motivate you to stick your head out the window or sit outdoors that little bit more than usual.

The RSPB hold their annual Big Garden Birdwatch every January, too. The world's largest garden wildlife survey, with 590,000 people taking part in 2025, it has been taking place since 1979 and has become a much-loved annual event that helps give the charity a valuable snapshot of how garden birds are doing in the UK. So, why not take part yourself and submit your garden records?

Winter staples

It's worth mentioning in this January section the winter patch activities that are covered in detail in other monthly sections (including November, December and February) – most of these are absolutely worth applying in January as well. They include routinely checking gull flocks and roosts, scrutinising deep bodies of water for diving waterbirds, counting roosts and seawatching.

02 February

February can be a mixed month for the patch birder. Winter birding will still offer plenty of excitement, but the novelty of a new year list might have begun to wear off, and the days are still on the short side and often filled with unenjoyable weather. There's plenty to keep you busy, though, such as checking wildfowl and gulls, and tracking down mixed flocks of seed-eaters, even if the idea of the spring's first Swallows and Wheatears can remain distant.

Furthermore, this month the monotony of winter birding does gradually give way to the wonderful first signs of spring and, on pleasant days (particularly in the second half of the month), many species are in song, with some even beginning to nest. If it's been a mild winter, which is an increasingly regular thing, then the frantic rush of plants growing and flowering weeks or even months ahead of schedule picks up significant pace during February. And even in a cold winter, on those warmer days the first butterflies might flit by, giving an uplifting jolt in the process. I also always think there's that feeling of having 'made it' through winter, with March heralding the first truly transitional period of the birding year.

Early breeders

Despite the continued winter vibes, with limited daylight, make no mistake – the breeding season is well underway. Plenty of resident species are in full song by February and mild, pleasant days can produce wonderful examples of the early spring dawn chorus, in other words the more regularly encountered British species singing in the mornings.

A fun thing to do in February is to make note of every species you hear singing for the first time. The composition will change almost week by

Logging the first time you hear different species singing is one of the pleasures of patch birding in the spring. Chaffinch often gets going in February.

Signs of breeding behaviour can be observed already, with some species building nests and others already sitting. This Long-tailed Tit is carrying nesting material.

week at this time. Two common species that often begin properly singing in February are Blackbird and Chaffinch, both of which are well worth taking the time to enjoy.

Indeed, everywhere you look there are signs of spring. In the garden, Robins will be firmly holding territories and Blue Tits will be paired up, perhaps prospecting boxes or engaged in their cute display flights. At farmland sites, rookeries will be bustling with noise and commotion, while Skylarks will be performing their first song flights. Meanwhile, Mute Swans (and various other wildfowl) and Great Crested Grebes will begin courtship routines at waterbodies. Making notes of breeding behaviour is important year-round and there is plenty to observe and document as early as February.

Some species are even deeper into the breeding cycle by February. Grey Herons and Ravens might already be on nests, with plenty of other birds, including other corvids and Mistle Thrushes, building them. The latter's nests can be fun to try to spot – they are large and untidy, often high up in the fork of a tree. Another species to try and note nest-building in February is Long-tailed Tit. Although not a particularly early breeder, they have such an elaborate nest that they start building it several weeks before egg-laying A nesting pair may take up to two weeks to build the oval structure of moss, lichen and spider webs, and then another one or two weeks to line it with up to 2,000 feathers.

We are only weeks away from the more classic harbingers of spring, such as Cuckoo and Sand Martin, so be sure to appreciate the more common species and subtle hints of this wonderful time of year, plenty of which are in full evidence already.

Tricky residents

A good way to fill winter time on patch, especially in the first few months of the year, is to try to target the trickier resident species in your area. During the first few weeks of the year, you may have focused on the more difficult winter visitors that will be gone come spring – and rightly so – but once most of these have been seen, and during periods of otherwise uninspiring weather, it's worth strategically targeting the less easy to see local resident birds.

Of course, what they constitute will vary depending on where you live, so local knowledge is key. Examples include owls, which for many patch birders are not always easy to locate. Little Owl is a good example. This is a bird that seems to have suffered a marked decline across much of England and is now tricky to see in some areas. February is a good time of year to stake out known or historic sites, with birds especially vocal as the breeding season commences. Barn Owls are often worth seeking out during the shorter winter days, too, with dusk at a much more agreeable time than during the summer months.

In February, try looking for more secretive species at dawn and dusk, such as owls and Water Rail (pictured).

Long-eared Owl is another species that can be best looked for in February, when males are singing well after dark. This is doubtless an under-recorded British species, and investment of time in identifying and then searching promising areas of habitat (coniferous woodland, hedgerows and shrubby thickets) may well pay off. Display commences at this time of year and is an enthralling sight. Males patrol their territory at dusk and dawn, stalling to clap wings below the body, and sometimes giving a 'booing' call as they go. Copulation sometimes follows these displays. Even if you don't find a territory, you might stumble across a roosting bird.

Several woodland species may also fall under the 'tricky resident' bracket, depending where you live. Lesser Spotted Woodpecker (see pages 76–77) and Goshawk (see pages 78–79) are two examples, along with Hawfinch, Firecrest, and Marsh and Willow Tits. Indeed, February can be a fun month to explore your local woodland sites. As discussed earlier, plenty of common species will now be in song and such places can be surprisingly full of life. Wetlands are worth checking for elusive resident species, such as Water Rail and Bearded Tit.

Little Owl is worth looking for in late winter and early spring when territorial birds are vocal.

Patience on a fine early spring morning may reward you with a local Goshawk displaying over a woodland.

Displaying raptors

February will usually produce the first proper spring-like days of the year, when the temperature may push double digits Celsius and the sky is big and blue. Such conditions are absolutely ripe for raptors to get up on the thermals and display. This is a particularly peaceful thing to take in on patch: simply find a slightly raised area then sit and watch. Tumbling Buzzards and slow wing-beating Red Kites are a joy to watch, their frequently accompanying vocalisations always carrying a sense of wilder places. The under-appreciated display of Kestrel can also be enjoyed at this time of year, with males flying up high to allow their underwings to shimmer in the early spring sunshine.

February, and indeed March, are also optimum periods to look for displaying Sparrowhawk and Goshawk. The reason that early spring is prime time is due to their display periods. Goshawk's is drawn out – territorial activity can begin as early as November and runs through to the end of March, but there is a clear peak in February and March. Meanwhile, Sparrowhawk's display is restricted to a shorter timeframe, typically between mid-February and April.

Goshawk is a particularly desirable bird and is increasing – and spreading – across much of Britain at a striking rate. Even if your patch or local area hasn't historically supported this species, it is worth bearing it in mind as we enter February. Various factors must be considered when seeking out Goshawk, including habitat, weather and time of day. In Britain, it's typically a forest and woodland species, with a strong preference for areas of mature conifer. Finding suitable habitat is one thing, but locating a good viewing area is another – and this is where an OS map (or some winter foot-soldiering) is useful.

Seek out a relatively high location, set back from the forest or at least looking over much of it. Goshawks tend to like hilly landscapes and this can be useful for finding a vantage point, as being 'above' the birds is always better than being below or at the same level (though not essential). Seeing them at this time of year can often be down to an interplay of three factors that only a Goshawk could truly explain: time of day, wind speed and visibility. In human terms, it helps if wind speed is at least a force three – this encourages raptors to become airborne and, in the case of Goshawk, perform their display flights. Naturally, visibility needs to be fairly good when staking out Goshawks due to the distances often involved, though 'classic' clear blue skies aren't essential.

On early spring days, these two factors tend to come together from about late morning. Once you're in position, it's a waiting game. Watching a Goshawk perform its display flight on an early spring morning is a genuine thrill – I will never tire of watching these brilliant raptors shooting up and plunging down during their 'skydancing' exhibitions. And even if you don't have Goshawks locally, time spent in the early spring sunshine watching Buzzards, Red Kites and Sparrowhawks is undoubtedly time well spent. And who knows what else might fly over!

Wader returns

The first migrant passerines usually appear on British shores in early March, but waders tend to start moving some weeks before this. Locally breeding species always arrive first, before the big push of species travelling greater distances, culminating in those heading to the high Arctic in May and even into July. Indeed, wader migration will be a strong theme in this part of the book!

One of the first signs of spring, especially for inland birders blessed with decent wetland habitat, is the return of certain wader species on breeding territories. Many arrive back during the opening days of February. Oystercatchers are often the first, and birds may even be on nesting grounds in January. Ringed Plovers tend to follow, with the second half of February seeing birds begin to move back from coastal wintering grounds.

Curlew is another species that likes to get back to its nesting sites early. Displaying birds are truly evocative, and you should count yourself lucky if this is a breeding bird in your area. Migrants are often heard before being seen, and February and March is the peak time for northbound birds. I'm very fortunate to still (just about) have Curlews breeding on one of my patches. Valentine's Day has traditionally been considered the marker from which to expect the first returning birds, though this has rarely been the case in recent years, with early to mid-March more typical.

Common Redshank and Avocet tend to return early to breeding grounds as well, even if March (and April, especially for Redshank) are the peak months.

Rinse and repeat

Despite all the above references to early spring signs, the reality is that much of February is wintry. The key is to persist with your patch – even if it feels like you've carefully gone through the local Tufted Duck flock or gull roost for weeks with no reward; the next time might be *the* time something rare is found.

Consistency (and perseverance) are both highly important factors when it comes to getting the most out of patch and local birding, so even at the end of a long winter, keep pushing yourself to get in the field and make the most of what you have in your area. After all, those brilliant spring days are within touching distance now.

The first returning breeding waders arrive at some sites during February, including Avocet.

03 March

I often find myself wishing I could press pause during March, which is perhaps my favourite month. It's such an uplifting time of year for naturalists, especially if you're a patch birder – any time outdoors is likely to yield a few signs of spring, be they subtle or great. After months of winter, the days are getting noticeably longer and things are finally leaping into life – patch sessions shift from repetition to anticipation amid a feeling of revitalisation. Yet March lacks the frenetic, expectant nature that April brings, allowing one to peacefully soak up the slow build-up of the year's most brilliant season.

March is the best time of the year to connect with the sadly declining Lesser Spotted Woodpecker.

As a result, there is plenty to get your teeth sunk into during this month. Many patch birders will agree that March signals the time of year when local efforts properly ramp up, with spring migrants beginning to arrive and the first proper goodies of the year conceivable, certainly in the second half of the month. And this is all set against the last days of winter playing out, meaning that a whole range of patch possibilities are on the cards.

If you go down to the woods...

Early spring is a wonderful time to enjoy woodland birding. Wherever you live in Britain, you should have some of this habitat nearby, even if it's marginal. Most resident species will be singing now and the dawn chorus can be properly enjoyed for the first time – it is worth appreciating the more routine songsters before the morning soundscape is supplemented by various summer migrants. Trees and herbaceous plants are beginning to flower, and woodlands can provide a real sense of positivity.

March is also the peak time for drumming woodpeckers. In Britain, there are only two species that properly drum: Great Spotted and Lesser Spotted Woodpeckers. The latter has suffered a sorry decline across the country and is now uncommon and localised – but it remains a tantalising, if tricky, patch target if you have suitable habitat.

At this time of year, with no leaves on the trees, the species is both audible and visible – something they very rarely are during the rest of the year. Several factors render finding 'Lesserspots' challenging. They are generally elusive birds, and have declined massively in most areas; they also have huge ranges outside the breeding season (hundreds of hectares). This generally shrinks at this time of year, though, as they settle down to breed, but by mid-April they can once more become incredibly tough to locate.

Woodlands are a joy to bird in early spring, with lots of breeding activity taking place after the long winter.

Pretty much any area with trees can hold the species – from relatively open parkland to extensive mature woodland – but generally areas with high levels of dead wood or woods associated with wetlands are the most productive. Within these areas, they are often found on the edges rather than the deepest parts of the woodland. Lesser Spotted Woodpeckers tend to be quite faithful to a site for many years, but they can move around within these ranges a lot. So, it's worth visiting areas the species was once known to frequent, even if it hasn't been seen there for some time.

Early mornings are, unsurprisingly, best, with drumming and vocalisations more often than not at their peak at this time. Getting your ear around the drumming is very useful – it's softer and flatter than that of Great Spotted Woodpecker and also lacks a flourish at the end, but does have a more intense, 'machine gun' quality with shorter intervals.

Time spent in woodland at this time of year may lead you towards other rare breeding birds. Hawfinch is an obvious example. By March, many birds will still be in winter flocks, but males will also begin singing. Use recordings to remind yourself of their subtle song before heading out.

Marsh and Willow Tits are also singing readily at this time of year and, in some areas, both are good patch birds, particularly Willow, which, as Britain's fastest declining resident bird has been lost from many areas, especially in southern England. Firecrest is another desirable local bird that begins to sing with gusto in March, as the resident population is supplemented by returning migrants, and it too is worth seeking out during early spring woodland wanders.

Unsung migrants

March, of course, heralds the initial return of many of our true and most celebrated summer migrants – and there will be more on them later. However, there is a whole range of subtler, less-appreciated migrants that move during March, some of which produce impressive spectacles. These unsung migrants offer a wonderful preliminary taster of the big migration push to come and are well worth appreciating in their own right.

Vis-mig can be worth undertaking during March. This key element of patch birding is covered in greater detail on pages 130–132 and 138–140, but in early spring evidence of movement can be enjoyed overhead in the right conditions. Many finches head back to Scandinavia during this time and species

Meadow Pipit passage is an unsung but notable element of spring birding, with late March and early April often producing the biggest days of movement.

like Chaffinch and Goldfinch can be detected in large numbers. Coastal sites typically score the biggest tallies – such as the 9,000 Chaffinches seen heading east at Dungeness, Kent, in March 2010 – but the spectacle can be witnessed inland, too.

Starling is another species that heads back to the continent during early spring. It may not always be recognised as a migrant, due to its year-round presence in Britain, but in the winter our resident birds are joined by massive numbers of continental individuals. It's not unusual to look up during March and catch a glimpse of a large, tight flock heading in an easterly direction. It's certainly a theme in my area, and I always ponder the origins and destinations of these high-flying groups.

Perhaps a more detectable overhead March migrant is Meadow Pipit, helped by the frequency with which birds call while on the move. For me, this is one of the first proper signs of spring movement. Typically occurring in the second half of the month, there is normally at least one big day per year, when it seems as though Meadow Pipits are constantly calling overhead. Birds heading back from European or southern British wintering grounds to northern breeding sites can be experienced anywhere. The diurnal nature of these movements means they can be easily enjoyed – and there's every chance you could score a bonus Merlin in tow, heading in a similar direction, or pick out a vocalising Rock or Water Pipit. On big 'Mipit' days, if time allows, it's well worth sitting and seeing how many birds you can count moving north.

Pleasant March days may see Mediterranean Gulls migrating overland towards the continent.

You can also see larger birds moving overhead in March. Having likely been a notable winter presence wherever your patch may be, many gulls begin to head back to breeding grounds at this time. Groups of Common and Black-headed Gulls can be seen sometimes at great height; in my area, there is always a marked cut-off of the former species locally, which goes from being one of the more conspicuous winter visitors to totally absent during the second half of March. Lesser Black-backed Gull is another species perhaps overlooked as a migrant, but many are long-distance travellers and March sees them move between winter and summer grounds. Often I won't see many (or any) Lesser Black-backed Gulls until late February and early March. For some inland patch birders, myself included, Great Black-backed Gull is a high-value prize – and March is one of the best months to hope for a flyover, as birds cut overland during transit from coastal areas.

More unusual gull species are possible at this time, too – think Kittiwake (especially during or after rough weather) and Mediterranean Gull.

With increasing numbers being seen in England, Cattle Egret is a species worth bearing in mind during March.

Indeed, a rather recent phenomenon, at least in the south-east of England, has been the overland migration of Mediterranean Gulls in March and April. This is probably due to a burgeoning winter population in the south and west (Cornwall, Dorset, southern Ireland and Wales) that is swollen by birds from the continent (many British ringing recoveries are from the Low Countries and Poland). Presumably, come the spring, many migrate back east over southern Britain, and their broody feelings mean they're especially vocal, rendering them detectable as they pass over at height.

Another recent March migration element to consider is the return passage of egrets. Both Cattle and Great Egret numbers have shot up in England during the last decade, with strongholds in the West Country. March seems to be a good time to connect with migrating birds heading back west, having presumably dispersed for the winter to the east of their breeding grounds. On the other hand, it's thought that some of our wintering Bitterns arrive from the continent. March sees birds head back and, on still, clear nights, many decide to make that return flight. If you have Bitterns wintering in your local area, trying to target a departure flight is a fun thing to do and you might be treated to the curious, Great Black-backed Gull-like flight call as they take off from a reedbed into the night sky. In fact it's worth taking a punt on this species in any area of promising, perhaps impenetrable habitat – you might be rewarded with a surprise Bittern record.

Some less-celebrated local songbird migrants may also stop off. Spring Stonechat passage peaks during early March, and breeding sites devoid of birds during the winter may suddenly become occupied, or a scrubby field that doesn't usually hold any may support several for one or two days. All these birds may not be as coveted as the first Swallow or Wheatear of the year, but they all contribute a sense of spring's arrival – and few feelings are as enjoyable in the natural world.

Early migrants

While a morning of significant Meadow Pipit movement or a northerly gull passage should be relished, there are few feelings in the patch birder's calendar as uplifting as that of seeing or hearing the first trans-Saharan spring migrant. It's a month when most birders excitedly await their first sighting of a Wheatear hopping across a grassy field, or a Sand Martin zipping freely above a waterbody. We persevere through the cold and grey winter months knowing that these spirit-lifters will eventually

White Wagtail (left) is one of the first spring migrant passerines to arrive and is often found amid Pied Wagtails (right).

return and render patient March birding – and indeed the winter slog – worthwhile.

The list of classic, expected summer migrants is a long one, but those that should be realistically expected in March is fairly short. Of course, our warming planet means many species are arriving earlier and earlier – it's not unusual now to see birds normally associated with April returns appearing back in March. However, the classic suite of March migrants remains a good guide with regards to species to target in this month. Weather plays a key role, of course, and if southerlies are blowing then migrants will arrive.

There is a migrant that often reaches our shores a little before the much-vaunted Wheatears and hirundines: the subtly beautiful White Wagtail. As a subspecies, White Wagtail tends to be slightly neglected and forgotten amid the jolly jamboree of spring arrivals. But an encounter with one of these suave-looking birds is always a treat. Indeed, White Wagtail may well be the first 'true' spring migrant you clap your eyes on each year and for that reason alone it is worthy of celebration.

For many of us, the first trans-Saharan spring migrant we will see is Sand Martin. These are always the first hirundines to arrive en masse and, by the end of March, plenty will have added this species to their year lists. Increasingly, the first of the year are arriving in February, but early and mid-March remains the peak arrival time. Mild southerlies, perhaps combined with low cloud or rain, may produce the first flitting over a waterbody – a truly uplifting sight. The first Swallows will often arrive in March, too, though typically later than Sand Martin.

Waterbodies may also produce the first Little Ringed Plovers of the year. Another species that is returning increasingly early, Little Ringed Plovers aren't necessarily common, but they are gettable for most patch birders in Britain (especially in England). The first birds may appear at a similar time to Sand Martins in early and mid-March; by the end of the month, pairs might be back on territory.

Every patch birder is excited to see their first Wheatear of the year.

Up there with the first hirundines of the year in terms of symbolism is Wheatear. This is perhaps one of the most iconic patch birds. Given that it typically breeds in upland areas, Wheatear is strictly a passage species for many of us – even if it can be common during these seasons. Clapping eyes upon the first of the year, typically in the second half of March, is always a memorable occasion.

Many other migrants can make a return before March is out, though for most of these species the peak arrivals occur in April. Still, it would be remiss to not mention two common warbler species: Common Chiffchaff and Blackcap, the familiar songs of which are likely to be heard before the month's out. Common Chiffchaff in particular may arrive en masse if suitable weather occurs in the first half of March, and suddenly an area previously devoid of this species may be filled with vocal males settling into their breeding territories. Be sure to appreciate singing Chiffchaffs and Blackcaps before they become part of the soundscape furniture.

Waterbird shifting

Waterbodies will experience significant turnover during March. Some species usually regarded as resident, such as Coot and Moorhen, shuffle around at this time, along with grebes, with Great Crested Grebe returning from coastal wintering grounds to fill up summer breeding sites. At my wetland patches, I can visit on one day and make a good count of wintering Coots, and then return the next day and see barely a quarter as many. Similarly, a site that's barely held a Great Crested Grebe all winter might one morning be dotted with several summer-plumaged adults that are ready to breed. With all this movement underway as birds return north, March represents an excellent time to get lucky with a scarcer grebe species – particularly Black-necked – or even an inland diver.

March is a busy month for many duck species, too. With wildfowl on the move, heading north for the summer, now is the ideal time to check out any local flocks, combing through the regular species for something scarcer. In the winter months we have discussed the importance of cold weather with regard to waterbird movements. However, often – and increasingly – such conditions don't come to pass, and thus wintering duck populations remain somewhat stationary. However, with March's longer days and warmer conditions, many begin to move – and this provides the opportunity to search for scarce and rare species. During March and April, Common Scoter undertake nocturnal overland migration across England. This movement involves large numbers of birds flying east and north-east, often from the northern Irish Sea and English Channel to the North Sea. This passage

March is a good time to check waterbirds for birds on the move, including Black-necked Grebe.

varies annually based on weather and other factors and has been revealed in recent years due to nocturnal sound recording efforts and the increase in usage of thermal imaging cameras.

Watching the same site can reveal a constant turnover of birds, with the ever-present chance of a surprise in this month. Diving ducks are particularly worthy of scrutiny. Being quite common and widespread, two species – Tufted Duck and Pochard – are typically easy to locate, and you will know which sites in your area hold the most. For most patch birders, nobody is too far away from a suitable site. Diving ducks will be present at such sites all winter, but March movements mean that their numbers and distribution begin to change, adding an element of unpredictability and the potential for something really surprising.

The variety of birds can be high at this time of year, both in terms of numbers and composition, and their length of stay may be short. If possible, daily (or even more frequent) coverage might pay off. It is worth emphasising that any site, however small, could host something unexpected. Among the commoner possibilities are Goldeneye and Red-crested Pochard, along with Scaup. Rarer goodies are absolutely possible, though, with March a particularly good month for Ring-necked Duck. Lesser Scaup and Ferruginous Duck – both national rarities and representative of a major patch find – are well worth bearing in mind. Keep up to date with national bird news as well – if it's been a particularly good winter for one of the scarcer species, for example, then you might be in luck this month as birds move around.

Dabbling ducks are also on the move at this time, with common species like Eurasian Teal and Wigeon returning to breeding grounds. Pintail, for some, marks a good-value patch bird, but one species strongly associated with March – and worthy of special mention

Rare duck species such as Ring-necked Duck may be found amid *Aythya* flocks in spring.

here – is Garganey. Another iconic early spring migrant, this truly stunning duck is arriving in numbers during the second half of month. Occasionally, influxes – often driven by wet winters (and thus more habitat) and south or south-easterly winds – will mean many more than usual are around. They can be surprisingly unfussy with regard to where they turn up, though typically they prefer freshwater habitats, especially shallow, eutrophic wetlands. Generally, this means flooded grasslands, water meadows and ponds with sufficient riparian vegetation. Even if you expect Garganey annually on your patch, finding one always delivers a high level of satisfaction.

Carefully scrutinising gatherings of common wildfowl may yield rewards. In this photo a Green-winged Teal is tucked in amid Eurasian Teal and Wigeon.

04 April

April is a truly wonderful time to be out birding, wherever you are. Birdlife at this time of year is rich and varied, while the unpredictability of migration makes each day as exciting as the next. Following the inspiring, gradual spring build-up of March, April is full-on in terms of breeding and migration action. Turnover is daily, new birds appear back on territory en masse and every corner of your patch will seem to hold the possibility of a nice bird or two. The longer daylight hours facilitate the opportunity for pre- and post-work birding and, whether your patch is a coastal hotspot or an inland cityscape or somewhere in between, the reality is that migrating birds can turn up almost anywhere on passage.

At the same time, it can be a somewhat overwhelming time for the patch birder – just where do you begin with all of the migration taking place? Well, there are certainly a few themes and elements to bear in mind during April, and the key thing is to enjoy every moment.

Logging migrant arrivals

A dominant theme of April is the arrival of the vast majority of our migrant breeders. From Common Tern and Common Sandpiper to Tree Pipit and Yellow Wagtail, most summer migrants appear in good numbers during this month. New species for your year list may be added on a daily basis during good runs of weather for migrant arrivals. Every year, without fail, patch birders are delighted by their first encounters with these returning summer visitors.

The return of these species represents veritable milestones in the patch birder's year, regardless of whether or not you're keeping a sharp eye on your year list. After a long winter, their appearance is reliable – but the timing and pattern of their arrivals varies annually as numerous factors influence migration. Every spring is different: one species may be first seen especially early one year, but then later than expected the following year.

Weather conditions play their part, with a run of mild south-westerlies sometimes causing mass arrivals well ahead of the expected dates. On the other hand, chilly north-easterlies can have the opposite effect. Furthermore, where you're based will dictate how early or late you might expect to see your first Cuckoo or Garden Warbler, for example. If you're in southern England, you could well be enjoying such species many days or even weeks earlier than if you're based in northern Scotland.

A really good habit to get into is logging your bird records and, by default, the first yearly records of summer migrants (see page 56–57). It

Common migrants, such as Common Sandpiper, arrive en masse during April.

The return of local Swallows is eagerly anticipated by patch birders each year.

is worth creating a separate list of your personal early and late dates of each common (or expected) migrant. It is fascinating to compare with previous years and, over the course of time, you'll probably notice earlier arrival dates for many birds. It is well known that climate change is driving the earlier arrival of migrant bird species in spring. Take Swallow, for example. In the UK, the average arrival date of this iconic summer visitor has advanced by more than two weeks since the 1970s. For Sand Martin, the figure is as high as 25 days. Keeping your own personal records – and ideally sharing them with a citizen science platform – will help contribute to the overall picture of these changes.

As mentioned, every year is different. Normally there is a species or two that has a particularly good or bad spring. This can be dictated by a range of factors, including how successful (or not) the previous breeding season was. In 2024, my home county of Surrey experienced a notably strong April for Grasshopper Warbler. This species is rather rare in the county now with only a handful of records in a typical year (of which just one or two are in April). In April 2024, however, double figures were recorded. Keeping an eye out for trends like this will boost your knowledge of what to target on your own patch, especially if it involves a migrant you usually struggle to connect with in a typical year.

Continental overshoots

For the rarity-finder, spring is synonymous with overshoots. These are species that migrate north to the southern European continent, but are generally rare or scarce in Britain. Each spring, when conditions play out, some individuals migrate further north than their usual range and reach our shores. While working through the daily turnover of common migrants (and logging their arrivals), it pays to keep in mind the possibility of something rarer.

Keeping a sharp eye on weather charts, the direction of the wind and

Rare overshoots such as Hoopoe can arrive in April.

conditions in Europe (and even North Africa) is wise. Many overshoot species will have migrated north directly from Africa, where they have wintered, rather than from southern Europe. Therefore conditions that far south will have an influence on what could be found on your patch.

There are various elements of the weather to consider. Perhaps the most important is for the wind to be in the south or, to a lesser extent, the east. You don't want conditions to be too fine and sunny, however, as cloud, murk or precipitation are best for grounding northbound migrants. That said, a clear night to the south is better for nocturnal migration – when this meets a cloudy or wet start to the daylight hours, then things could be particularly productive. After all, if conditions aren't conducive for migration further south, then birds won't bother moving at all. Optimal April mornings for decent migrant arrivals are usually when there are clear conditions overnight on the near-Continent, transforming to misty or drizzly weather in southern Britain in the very early hours, grounding the migrants and creating a fall. Predicting such conditions is not always easy – and depending on where you live, conditions further south in Britain will have little impact – but a good, broad rule of thumb is that the less calm and clear the weather, the more likely it is that migrants will be grounded.

Migrating birds in spring are generally in far more of a rush than those in the autumn, as they need to claim breeding territory ahead of their rivals, and often their rest periods can be fleeting. So, it pays to be alert – and to tune in to the vocalisations of some species, which you may first detect flying over or singing momentarily from cover.

The chances of finding particular species is affected by which type of southerlies are in play. Prolonged southerly or south-westerly airflows originating from the Sahara, for example, are perfect for overshooting species like Hoopoe, Purple Heron, Golden Oriole and Serin. On the other hand, a more easterly airflow may produce migrants that are heading to Scandinavia, especially if you're based along the east coast, with Bluethroat and Wryneck two examples. Differing airflows have different advantages – and, of course, where you're situated will mean there are different levels of expectation – but on the assumption that a steady south-easterly, southerly or south-westerly is in play, it's likely that these more unusual species will be arriving alongside the commoner birds.

If you're based inland, it can sometimes feel like you're missing out on the overshoot front. However, it's well worth keeping note of which overshoot species are experiencing influxes in any given year. For example, if either the south or east coast of Britain has enjoyed a bumper number of April Hoopoes or Night Herons, then there's every chance many of these birds will continue inland, where they may be picked out by sharp-eyed observers in the days (and even weeks) after the main arrivals. Their appearance is less associated with immediate weather, of course, but keeping your finger on the pulse is simply another element of playing the conditions.

Habitat plays a crucial role in locating migrants, especially overshoots. While coastal sites are best positioned, especially headlands and marshes, there is plenty of scope for possibilities inland. High hilltops are natural points of fall for migrants, especially those with limited cover – 'vertical inland headlands' as they are dubbed by some birders. Some places at lower elevations, such as river valleys and large waterbodies, serve as funnels and magnets to migrating birds. Avoiding areas with dense vegetation is a good rule of thumb, too – there's no point making a difficult job even harder!

Of course, these discoveries are rare, even if you're lucky enough to have a coastal patch. Inland, such goodies are even less likely. Local birding is all about context, though, and April fall conditions are prime time for certain species that get the pulse racing, even if they aren't especially scarce. For example, it's an excellent time to bump into a Pied Flycatcher, Wood Warbler or Redstart – perhaps not overly notable on the coast, and for some maybe even an expected breeding bird. However, in many cases birds like this are representative of a major patch success and should be cherished.

So, in summary, keep a firm eye on the weather (especially the wind direction), try to avoid searching during clear, calm conditions, don't expend too much energy working large areas of cover, and get familiar with vocalisations.

Opposite page: Northbound migrants, like this Wood Warbler, may hold temporary territory on passage in April.

April showers

The proverb 'March winds and April showers bring forth May flowers' was first recorded in 1886 – but it is still relevant today. Showers are synonymous with April in Britain. Many of us will have experienced dramatic change in the weather on a single April day, with sunshine and warmth soon replaced by rain or even sleet and snow. Indeed, in London, the date with the smallest chance of dry weather is, on average, 27 April.

For the patch birder, April showers are there to be taken advantage of – especially if you have a waterbody in your area. There is perhaps no better

example of the usefulness of 'twitching the weather' (going birding based on the conditions) than heavy downpours during this month. Such conditions will ground birds that are migrating overland, forcing them to ditch down at a temporary bit of habitat. While waders are important to consider, especially during the second half of the month, they tend to peak at the end of the month and during May. April, however, is prime time for birds like Osprey, terns and gulls.

As a result, skywatching on a showery day – even away from a waterbody – can deliver surprises. At this time of year, with spring migration in full flow, take advantage of cloud and rain. Keep an eye on the real-time weather charts and indeed the skies outside. If it looks like a front of heavy rain is imminent, consider dashing out to your nearest waterbody. This is something I've practised countless times in Aprils gone by and, although you may often find nothing has dropped in during the inclement weather, there will be plenty of occasions when you'll be rewarded.

Osprey has increased markedly as a breeding British bird and, although individuals can be encountered on passage from mid-March (and indeed some are back on breeding grounds before April has even begun), now is the prime time to score one locally. Showery days will force birds to fly lower and maybe even call in at a lake or reservoir.

Terns might also be on the cards during such conditions. Sandwich Tern is a real possibility throughout the month in such weather, with Arctic and

An easterly airflow and poor weather in April are perfect for connecting with Little Gulls.

A drizzly April day is prime time to see if any migrating birds have dropped in to to your local waterbody.

Little Terns becoming options from the second half of April onwards. Black Terns are more associated with easterly winds (also in early May), but they too should be considered.

Little Gull is a classic patch birder's bird – one that never fails to raise a smile, even in areas where it is slightly more expected. During spring (with April the peak month), large numbers pass through on their way to and from breeding grounds in northern Europe. Birds take shortcuts across land so can routinely be found at inland waterbodies. East or south-east winds can produce good passage across southern England and the Midlands, especially if this airflow occurs in conjunction with April showers, which force birds to temporarily halt their movements. This particularly beautiful gull – the world's smallest species – is always one of the highlights of the spring, and a good reason to be diligent during showery weather.

As mentioned, waders are very much in play during such conditions too, with single birds or small flocks dropping onto shorelines to avoid flying in the worst of the weather. April is better than May for certain species and offers significantly more species diversity than March.

Among the early movers (including throughout March) are Iceland-breeding Black-tailed Godwits; passage for this species is protracted and they can pass through well into May. A classic April wader is Spotted Redshank, with the middle and end of the month the optimum time for this desirable species. Get familiar with the calls – they are often vocal – and be aware that birds can pitch down almost anywhere as they journey back to Scandinavian and Russian breeding grounds.

Greenshank is another vocal species that will begin to appear in mid-April (and continue well into May), with its conspicuous nature making it a realistic target both as a flyover or at a wetland site. It's worth appreciating the more routine species, too. The first half of the month will see the last Green Sandpipers head back to breeding grounds, and it's the peak month for Common Sandpiper, too. These two species might represent patch gold

in a dry area or at a particularly small waterbody.

In terms of passerines, rarities are not to be forgotten in such weather. One good example, especially towards the end of April, is Red-rumped Swallow. On wet days with large gatherings of hirundines (especially arriving House Martins), be sure to check through the masses carefully.

Seawatching

While wetland birding during April showers equates to a form of inland seawatching, some birders are lucky enough to have the coast in their local areas. Mid-April through to mid-May is one of the best times of year to invest patch time in seawatching, as birds head up the English Channel and up the west or east coasts to northern breeding grounds.

Diversity is perhaps at a year-round high, with wildfowl, grebes, divers, waders, gulls, terns and skuas all on the move. Knowing the best conditions for your local area is key. On big days, huge movements of species like Little Gull, Arctic and Black Terns, certain wader species like Whimbrel and Bar-tailed Godwit, and more, can be enjoyed.

Depending on where you live, skua passage might be a theme – and late April through to mid-May represents the best time to connect with Pomarine and Long-tailed Skuas in some areas (such as the south coast for the former species and the north-west for the latter).

Spring seawatching often involves lengthy stakeouts (especially given the long daylight hours) and patience, so make sure you come prepared. A comfy chair, a rain jacket or umbrella and a suitable supply of food and drink should all be considered!

Farmland scrutiny

While the excitement of waterbodies, overshoots and seawatching can dominate thoughts during April, it's worth paying some attention to your patches of farmland as well. After all, April is often the last chance for many weeks to scan fields before crops grow too high, and finch and bunting flocks may still gather in the early part of the month, with some species such as Linnet and Corn Bunting often lingering in groups later into the winter than most. The possibility of goodies like Stone Curlew or Dotterel shouldn't be dismissed either – the former has a knack of turning up on passage in stony fields (often including sheep grazing land), while the latter is associated with arable farmland (especially pea crops).

Paddocks are also worth scrutinising during April. At this time of year, far more so than during autumn, they are particularly attractive to migrating Ring Ouzels, which like probing the soft earth for earthworms and other food. Keep an eye on any paddocks

Above: Spring seawatching can produce a decent passage of skuas. **Opposite page:** On days of large arrivals of hirundines in spring it is worth being alert to the possibility of Red-rumped Swallow.

Ring Ouzel peaks in the spring in April. Check local horse paddocks for this species.

that hold decent numbers of Blackbirds and Mistle Thrushes (the latter may be feeding young in early and mid-April), or indeed lingering Fieldfares. You may just be treated to some of your best patch views of Ring Ouzel in the process, with smart spring males a particular treat.

April is also the best spring month for Redstart and Black Redstart. Both species breed in Britain (though Black Redstart is rare and localised), but encountering one on passage is always satisfying and, depending on where you live, may even represent a quality local find. The first glimpse of the orange 'start' (an old name for tail), 'shivering' in the habitual way of both species, is cause for excitement. Passage Black Redstarts move earlier, climaxing in the second half of March and first half of April, with the bulk of migrant Redstarts passing through during April and perhaps just into May. Both species are fond of paddocks for foraging, especially if sheltered hedgerows are nearby (particularly for Redstart), and can appear at any time of the day in spring.

Periods of easterly winds may produce arrivals of Black Redstart in April.

05 May

May can be an exhilarating month on patch. By now the days are truly long and, fingers crossed, the weather is warm and sunny. It can often feel like anything could turn up locally – especially if the conditions are right. It's a particularly good month for rarities and, during the first week or two, spring migration is perhaps at its most dynamic, with the last large volumes of passerine migrants arriving alongside a peak of wader movement, plus plenty more, from terns to raptors. As well as this, some of the last expected summer visitors arrive, rounding off the season of migration. When I was young, a local birder told me that the last week of April and first fortnight of May was the best period of the entire spring – to this day I think this is a reliable perspective and it's something I bear in mind annually.

On the other hand, May can occasionally feel like 'after the Lord Mayor's Show', especially in the second half of the month. After the lengthy build-up and anticipation, the bulk of migration is completed and action can be reduced to something of a trickle. In some years, that 'high summer lull' feeling may kick in before the month's end. Typically, though, even if there are signs of spring slowing down into a lazy summer, May will produce some brilliant patch moments – and it's important to not rest on your laurels, as this is a time of quality over quantity, with the rarest spring patch finds often towards the end of May and into June, involving birds that have come from furthest away.

Wonderful waders

Throughout these monthly sections, waders are a theme. Back in February we touched on the first species returning to breeding grounds symbolising one of the earliest signs of spring – and spring passage can run well into June. The northward push of shorebirds is truly exciting, whether you're an inland or coastal patch birder. Certainly, if you are land-locked, this group of birds offers a special sense of excitement, with your little area part of a bigger, pan-continental journey. Even areas with little water will be graced by wader migration – one of the reasons why it's so exciting!

As mentioned, the first signs of wader passage can be detected in February (or even late January), then it ramps up during March for several species. April is a great month for shorebird action and much of the content here can be applied to that month, especially its second half. However, it is May – particularly the opening week or two – that is generally regarded as *the* peak time for spring wader passage. In addition, at this time of year many species will be looking splendid in full breeding plumage – a far cry from the dull (and often distant) birds that adorn our coast during the winter.

Many migrating waders are anxious to get a move on to their northern breeding grounds and will only drop down to a suitable bit of habitat for a short period, to feed up and rest ahead of continuing their journey, or to avoid particularly inclement weather, like heavy rain or fog. This provides a special feeling of unpredictability to what you might encounter. Different species move at different times during the long spring migration window, too. As a result, knowing what to look for when is helpful.

Habitat is of course important. If you have a reservoir, coastal marsh or wetland in your area, then you're going to get waders. However, marginal habitat should not be ignored. River valleys, temporary gravel pits and flood meadows all have pulling power. It pays to be alert to the 'pop-up' wader habitats in your area – a particularly wet spring, for example, may produce a seasonal flood that otherwise wouldn't be there. And while water is important, it isn't essential. Many of us will have heard waders flying over at night as they migrate – species like Oystercatcher and Green Sandpiper are typical examples. And some waders can appear anywhere. I've seen Whimbrel on polo pitches and Dunlin in arable farmland before... so, keep an open mind, especially if the weather is inclement, forcing birds to fly lower or drop down.

On that note, it's important to pay attention to the all-important weather, in a similar vein to April's showery conditions, discussed previously. Warm days with sunshine and high cloud will mean many birds passing overhead at great height and out of view, but heavy cloud and, ideally, precipitation will cause birds logistical problems so if they do continue to fly they will do so at much lower altitude. That said, nothing is guaranteed during spring – quality waders will often appear at the most unexpected locations on the clearest of days, and rain doesn't necessarily translate to significant passage!

Longer sessions in the field will produce more results than quick checks. On a promising early May morning, for example, with cloud and the possibility of rain, or a marked headwind, consider staking out your best wader site – maybe even bring breakfast and a chair! Many birds will fly straight through, if they can, or just drop in for a minute or two. Some of my most memorable spring days on patch have been spent staking out my local reservoir and it's often the fleeting visitors that are the rarest. Consider checking more than once a day, too. If a morning's visit was quiet, that doesn't mean there may not be something special later in the day, especially during wet, showery

Wader passage can be lively during late April and early May. Whimbrel is a species symptomatic of this time of year.

Inclement weather and a north-easterly wind may produce inland records of Bar-tailed Godwit.

conditions. I've found inland Turnstone and Sanderling during May evenings (and sometimes in fine weather, too!).

Several species typify the most dynamic period of the end of April and first half of May. One of the classic examples is Bar-tailed Godwit. Always good value inland, late April and early May sees one or two big days for this species annually, and they can be surprisingly unfussy about where they turn up. Sometimes big flocks are involved and can produce quite a spectacle. With real luck, you may even connect with a flock migrating high above a dry landscape. A species with a similar peak to 'Barwit' is Whimbrel. Early May is prime time to connect with one and they can turn up anywhere, from flying over your garden to foraging in arable fields. They are one of the most 'gettable' passage waders.

Another classic May wader – and one of the best examples of a species looking far more wonderful in breeding plumage compared to winter – is Grey Plover. It is a rare prize inland but may drop in during rainy days. Wood Sandpiper is a classic patch birders bird, in a similar vein to Garganey and Little Gull, and its spring peak is also early May. Easterly winds are best for good numbers, though you are more likely to encounter one if you have a coastal patch, especially in eastern areas.

Dunlins move in large numbers during early May, too, and look fetching in summer plumage. Knowing the call is useful, especially if you aren't blessed with prime wader habitat, as this species will often move straight through if it can afford to. For some of us, with shorebird hotspots, Dunlin may be one of the more typically encountered spring passage species, with small flocks gathering at this time. These groups deserve scrutiny – Dunlin is a valuable carrier species, with even small flocks sometimes holding scarcer small sandpipers among them. They often travel with Ringed Plovers of the Arctic-bound *tundrae* subspecies, too.

While a familiar sight at coastal locations, an inland Turnstone is a real prize.

I mentioned Turnstone and Sanderling earlier. These are two species that we typically think of as strictly coastal, yet May is the optimum time for inland records. Breeding-plumaged individuals showing well on a reservoir or lake can almost feel like new species, such is the difference to encounters with them in duller plumages on a cold winter's day on the coast. Both are scarce inland, but May is the best month in the calendar year to find them at your local site.

Even rarer is Temminck's Stint – a classic early May bird. Numbers can vary year on year, and typically it's east coast sites that score the most records, but they are possible inland, especially when the wind is in the east. Bound for Arctic breeding grounds, finding one of these would represent a great patch find anywhere. They tend to prefer muddier or grassier locales than, say, a concrete reservoir. Other unlikely – but possible – May waders including Knot, Little Stint and Curlew Sandpiper.

And, of course, there's always the chance of something truly mega. Red-necked Phalarope is a late migrant, and the end of May (and early June) sees birds occasionally drop in at freshwater sites. Broad-billed and Marsh Sandpipers are other classic rarities of late spring. These are, of course, not to be expected, but both represent once-in-a-lifetime patch possibilities for the keen wader watcher!

When it comes to patch birding and waders, context is important. If you live on the coast, or near a large inland wetland, then your expectations and hopes will be far loftier than someone who is based in a largely dry inland area. I fit into the latter category – a single Redshank or Dunlin is a win for me! – and it's important to appreciate a species's local status in such cases.

Look for confirmed breeding

By May, many species will have had first broods – some may even have finished breeding for the year. As a result, it's a good month for confirming breeding attempts and, better still, marking down successful breeding. Many breeding records are of singing males or other

Big days

A 'big day' is a well-established birding concept. In short, a birder – or small team of birders – records as many birds as possible within 24 hours in a given area. Some big days are undertaken on county or even national level, with others confined to a patch or local area. Challenging yourself to a big day on your patch or local area is great fun – and early May is probably the best time to do one!

Sometimes big days are undertaken by several different teams, all competing to try and achieve the highest species total – this is known as a bird race. Often, though, individuals just want to challenge themselves to see how many species they can detect in their area. How much effort you want to put in is completely up to you. Starting in the early morning and finishing late afternoon, with a few breaks in between, can yield a lengthy list. On the other hand, some people like to begin at midnight on the dot, looking and listening for nocturnal birds (including migrants), and finish long after the sun has set!

You can make your own rules, but if you're competing in a team then typically at least two members need to hear or see a bird to count it. Part of the fun of big days is the planning and groundwork beforehand – working out exactly where to go for certain species. You'll find yourself making extra note of that Bullfinch territory or Sparrowhawk nest. There are always a few misses on big days, often comical ones involving common species; I've been on big days that have only just scraped birds like Nuthatch and Mistle Thrush. A successful big day is in many ways a reflection of your patch knowledge. You will know how many species are possible and where the best places are for some of the trickier ones.

The unobtrusive Bullfinch can be a tricky big day bird.

Early May is ideal for a big day as virtually all summer migrants are back on breeding grounds, but passage is still underway, so you could score a bonus Wheatear or Common Sandpiper (if these don't breed on your patch), for example. It's also a great time for something unexpected – I recall being part of a team competing in a race in different part of Surrey to my own patch when a Sandwich Tern was found by another group, a significantly rare bird for the area!

It's certainly interesting to work out – and try to achieve – the maximum number of birds you could see in your area in a single day. You might be surprised at the diversity your area offers. Of course, you don't have to do a big day in May. Some birders prefer late April, as they believe it produces a greater diversity of migrants. Others prefer the start of January, to help get their year list underway, and others fancy September as it offers a taste of both summer and autumn.

courtship, so indicate only potential breeding. Clearer-cut evidence, like the presence of recently fledged young, is very valuable. Even for common species this is precious data, so be sure to record it in your notes.

Wherever you live, there's a chance that rare breeders can be found in your area. The list of officially rare breeding birds is decided by the Rare Breeding Birds Panel (RBBP) and can be found on their website (rbbp.org.uk), along with tips and advice on how to look for and monitor rare breeding birds without disturbing them.

Monitoring rare breeding birds can be really satisfying, if you have the time to do so. If you're in southern areas, May is a brilliant time to scan woodlands for Honey Buzzard, for example, or listen for purring Turtle Doves in areas of suitable countryside. Meanwhile, in northern locales, nesting waders and seabirds can be monitored, as well as some raptor species and Ring Ouzel. Some species we may not think of as rare fall under the 'rare breeding bird' bracket, too, including Shoveler and Pochard. So, if you have these species on your patch into late April, be mindful that they might stick around to breed.

It's not purely about the RBBP-level (see page 57) breeding species either. Breeding evidence for birds that are in decline nationally but not yet monitored by the panel, such as Spotted Flycatcher, is well worth seeking out.

Enter the rare zone

In the April section we discussed overshoots, and these are worth considering throughout May as well. However, May has better form for delivering other rarities too – particularly during the middle and end of the month (often, and increasingly so, into June). Even though mass migration has virtually finished by the second half of the month, that needle-in-a-haystack mega is possible, particularly if winds are blowing from the east.

Turtle Dove has become a very rare breeding bird, so documenting a spring individual is important.

Periods of easterly winds in May can produce vagrants from eastern Europe and Asia, such as Red-footed Falcon.

The key is simply to stay alert and focused, even if it feels like summer is slipping into view. There are no cut-and-dry rules for locating late spring rarities – luck plays a huge part, but the main thing is to remain tuned in and optimistic. Time of day can be an important factor. For a singing Golden Oriole or Melodious Warbler, early morning is best. Shrikes tend to become most active once the day has warmed up a bit, while raptors – including Red-footed Falcon and Honey Buzzard – are best sought out during the middle of the day.

Dusk and beyond is an increasingly good time to spend looking for rarities, particularly if you're blessed with a wetland. Herons, including Black-crowned Night Heron and Purple Heron, may fly from a hidden spot close to sunset. Reedbeds and rank vegetation near water could yield a range of rare warblers, with Blyth's Reed, Great Reed, Marsh and Savi's all possible if you are very lucky. Once the sun has properly set, if you're feeling ambitious, you could listen out for crakes. Spotted Crake is a localised yet under-recorded breeding bird in Britain, and Baillon's or Little Crakes are not completely out of the question!

Last arrivals and thinking ahead

Although we're not even halfway through the year, the reality is that May signals the last of the arriving summer visitors – and if you're year listing, the bulk of your total will be achieved by now. For me, there are only three annually expected breeding species that I don't expect to add until May: Nightjar, Honey Buzzard and Spotted Flycatcher (though occasionally I won't see my first Hobby until May). By mid-month, it's likely all expected summer migrants will be in.

As a result, it's a time to reflect, especially if you're keeping a year list. The foundations for the rest of your birding year have been set and many species you add to your list hereafter will be hard-won. If you've chalked off all the expected migrants, you can now dedicate more time to seeking out the unexpected (and focus on studying breeding birds). It might even be an idea to write down some of the outstanding targets, including possibilities for the quieter June that lies ahead. You can try and establish what other species you can work to try and see during this steadier period.

06 June

Having commenced in subtle ways since as far back as February, spring migration finally winds down to an end in June. This is often considered to be a quiet month in birding terms, as things slip into 'high summer mode' – birdsong diminishes, spring passage stops and the days are long and usually pleasant. For some of us, attention may turn to mammals, reptiles and amphibians, insects or wildflowers – or simply some rest! Marking the halfway point of the year, I often find June – especially the second half of the month – to be a good time to unwind and recharge my patch batteries. The full-on, frenetic nature of peak spring can leave you feeling a little exhausted by now, especially if you've been committed to the earlier starts that the longer days slowly challenge you to meet as the year develops.

Despite this, the reality is migration never stops for long – birds are still heading north in the early period of June and, by the end of the month, the first return wader passage is evident. It's also a fantastic time to monitor breeding birds. Indeed, many of our resident species are nearing the end of their breeding seasons by late June and the countryside can be teeming with fledglings of all sorts of species. I've increasingly learned that it's important to stay on your patch toes in June, even if you can generally afford to dial down effort and take things a little bit easier.

Wader migration seemingly ever ends. Sanderling can still be on the move north into early June.

Shorebirds never stop

Waders never really stop moving, and migration for several species – chiefly birds heading to the high Arctic – can sometimes continue well into June. Commoner species, such as Dunlin and Ringed Plover, are typical examples of birds that can still drop in on passage at this time of year. Most migrant Ringed Plovers in June will be of the subspecies *tundrae*, which breeds in Arctic Scandinavia and Asiatic Russia. They look particularly smart by now and, if Ringed Plover is a good-value local bird for you, it can be a decent time of year to seek them out.

Ringed Plover and Dunlin are decent carrier species for rarer birds, too. Other high Arctic-bound birds – including Sanderling, Turnstone, Knot, Curlew Sandpiper and Little Stint – may be found tagging along with a group of commoner birds. The two last-named mark a nice find anywhere you live, while Sanderling, Turnstone and Knot are real inland goodies. In fact, late May and early June are perhaps the best times of the year to seek out inland Sanderling. So, if you've got a nearby reservoir or gravel pit, it's well worth seeking them out, especially if there is unseasonal wet or grey weather. Sanderling are great to watch when they're skipping along a sandy beach in their cold-toned winter plumage, but look quite different (and no less delightful) in late spring and summer, with breeding-plumaged birds a variety of different warm reddish colours.

Rarer waders are of course possible, too. Red-necked Phalarope is a bit

A breeding plumage Red-necked Phalarope would represent a major find for any patch birder in late spring or early summer.

Above: Rare waders are possible in June. Here, birders twitch a Lesser Yellowlegs found on an inland reservoir near London.

of a June speciality. Although a rare passage bird anywhere in Britain, it is nevertheless worth keeping in mind. Broad-billed Sandpiper is a full-fat BBRC rarity but June is a good time for this curious species and there are plenty of records from this month, both inland and at coastal sites. One tucked into a flock of northbound Dunlins is perhaps the most likely finding scenario. Increasingly, Black-winged Stilt is turning up in Britain, and June is an excellent month for the species, which can turn up almost anywhere – I recall an incredible record from 2022 of a female turning up on a small pool in the middle of a housing estate in land-locked Buckinghamshire!

So, even if it might seem like summertime and waders are not in play, it's worth checking – even in seemingly unsuitable weather – for shorebirds dropping in. I remember watching a glorious breeding-plumaged Lesser Yellowlegs on a Surrey reservoir (sadly not on my patch!) in early June 2022, which was seemingly travelling with a single Ringed Plover, in fine, clear early-summer weather. The next day it was photographed in Scotland!

And before you know it, return migration is underway. As soon as we pass the solstice it can feel as though autumn is beginning in some years, as early or failed breeding waders begin moving back south. Green Sandpiper is a classic species that will start appearing in suitable wetland areas from the second half of June onwards – sometimes even earlier in the month! GPS-tagging studies have shown that some individuals of this species stay on Scandinavian breeding grounds only for a matter of weeks, remarkably enough. Other species, often including British breeding species from further north, like Oystercatcher and Redshank, will also start migrating back.

Easterly excitement and southerly surprises

The birders' saying that 'the big one travels alone' is especially relevant for June. Generally, passerine migration is slow to non-existent in June – but, with luck, you could find a major prize. The opening week or two of the month can be a good period to keep an eye on both the weather and national bird news. If southerlies or easterlies – ideally warm south-easterlies – are in play then it can be a time for overshooting species to arrive in Britain, thus presenting good opportunities for a major patch find.

Classic overshoot species in early June include Bee-eater, Golden Oriole

Not long after the last spring migrants have arrived, the first southbound returnees can be seen, with Green Sandpiper a classic example.

Early June can be surprisingly good for turning up rare birds, like Red-backed Shrike.

Rare herons should be worth keeping on your summer radar, like Black-crowned Night Heron.

and Red-backed Shrike. A good thing about this trio of species is that they can turn up almost anywhere vaguely suitable. A typical Bee-eater find in Britain will involve a flyover, so learning the distinctive flight call is handy to say the least. Your ears will also almost certainly be involved in locating any Golden Oriole, with its evocative fluty song sometimes all you'll manage in an encounter with this species. Any wooded area may harbour a bird, even inland. Red-backed Shrike will be relatively unfussy on the coast, as long as there is a supply of suitable prey, though inland commons and scrubby farmland are also worth searching.

Two rare *Acrocephalus* warblers are possible in June off the back of south-easterly winds: Blyth's Reed and Marsh Warblers. The former is much rarer, but in recent times has shown an increase in British records, including in the spring. If you're situated on the east coast you are in the best position by far. Marsh Warbler will turn up inland – it is a very rare breeding bird so, should you have the good fortune of finding one, consider whether it's wise to share news publicly. Both species are fairly unfussy with their habitat choice but wetland margins are generally best; get your ear tuned into to the similar but nonetheless identifiably different songs of these two species. These two warblers – and indeed Bee-eater, Golden Oriole and Red-backed Shrike – are all prone to occasional bumper springs and it can pay to have your finger on the pulse in this regard. Keeping an eye on national bird news services may show that certain birds are in play and thus worth looking for locally.

June is also a good time for rare herons from southern Europe, especially during periods of drought in Iberia and France – a grimly increasing occurrence in the modern day. Species such as Purple Heron and Black-crowned Night Heron are fair game these days and well worth bearing in mind if your patch or local area has extensive wetlands or reedbeds. These species can often be elusive, mind, and dawn or dusk visits are recommended.

You can't discuss scarcer June birds without mentioning Quail. Some of you may be lucky enough to have this species as an expected local bird but that's not the case for many of us, including me! Arable farmland is the best place to listen for singing birds, ideally on a warm, still morning or evening. Grassland and meadows are good places to try too. In a 'quail year' hundreds of birds may reach Britain, with other years being poor. Again, it's worth tuning into the national birding news to try to get an understanding of numbers.

When it comes to June patch rarities, it's important to remember that patience is required. Often there will be little other evidence of migrating birds and it will often seem like you're flogging a dead horse. Exciting finds can be thin on the ground at this time of year but, remember, the big one often travels alone...

Those long summer nights

June offers the longest daylight hours – and marrying up your patch birding with the long evenings is worthwhile during this time of the year. On warm, pleasant evenings in June, a patch visit can take on a most tranquil feel. Sometimes it's nice to just head to a favoured spot then simply sit and watch. Of course, certain species can be targeted at such times as well.

Owls, if they have successfully hatched young, will be hunting as much as possible to feed their hungry chicks so are more visible than usual. I find June is a particularly good time to connect with my local Barn and Little Owls, for example. If you have suitable habitat for Long-eared Owls, go listening for the 'squeaky gate' call of juveniles at this time of year – such a discovery would mark a fantastic breeding record too.

If you live in an area where Nightjars occur, then June is the prime time to

Warm summer nights are perfect for looking for Nightjar, should you be lucky enough to have them in your area.

Woodcock is another species that can be enjoyed on summer evenings.

Species such as Common Tern are in full breeding swing during June.

enjoy their churring performances. An encounter with this species never fails to amaze – Nightjars are really curious birds that you can spend time with until the small hours of the morning. This is a species that is slowly pushing beyond the edges of its established range, so even if they aren't usually in your area, perhaps try some suitable-looking habitat.

Roding Woodcocks also have a mystique about them and perform particularly well on warm summer evenings, uttering their curious calls and flying around on broad, slow-flapping wings. Other waders too can be more active at such times, such as drumming Snipe, and Cuckoos will sing long after dark if they're in the mood.

Breeding codes

It's not all about rarities in June. And although it's important to stay alert and plugged in to possible patch rewards, it will be one of the quieter months of the year for unusual birds, for the most part. On the other hand, this is a great time to appreciate the commoner breeding birds, many of which are in the middle of their breeding season by now – or even coming to the end of it. Indeed, the first Cuckoos and Nightingales will already be heading south by the end of the month.

A visit to a woodland in June can be most enjoyable and serene – and teeming with fledglings of various species. It is simply uplifting to be surrounded by so much evidence of life and it's also a good opportunity to log as much confirmed breeding as possible, be it recently fledged juveniles or adults carrying food. Tits and common warblers make up most of the young birds roving around, as well as thrushes, Wrens, Dunnocks and Robins.

June is a great month to try and document rare breeding birds, like this Hawfinch feeding its chick.

And it doesn't need to be a woodland – literally anywhere in the countryside will yield signs of breeding success in June and all records logged are of value, so be sure to learn and use the right breeding codes if you're using BirdTrack or eBird. When you submit a checklist, challenge yourself to add a breeding code for every species you see (if valid, of course) – common or not. Volunteering to take part in breeding bird surveys, be it national or county level ones, is a great way to contribute treasured data as well.

Rare nesters

As well as the common species, June, like May, is a good month to monitor any rare breeding birds you might have in your area. Discretion is important – for many species it is sensible to keep your records private and share them only with your county recorder or the RBBP.

Try to work out which species may be worth looking for based on where you live. If you're in the south and have wetlands nearby, perhaps check local heronries for rarer egrets and herons. Or if you're in a heavily wooded area, staking out Hawfinch and Hobby might be a shout. In northern Scotland, it may be that breeding waders like Greenshank and Wood Sandpiper are possible. Even if you live in an urban area, species like Peregrine and Black Redstart could be

Studying rare breeding birds, either on a local or national level, can be immensely rewarding. This Curlew chick was born at one of few sites the species breeds at in southern England.

on the cards and would represent an excellent breeding find.

It's not just RBBP species to look for either. If you can obtain records of species that are rare breeders on a local or county level, or suffering national declines, then that's a job well done. Breeding data of species that are declining, like Lapwing, Linnet or Curlew, are really important.

Similarly, there may be nationally common species that are particularly rare breeding birds in your county or local area. For me, examples include Teal, Meadow Pipit and Redpoll – species that may be commonly encountered in the breeding season elsewhere in Britain, but are very localised and notable in Surrey. Engaging with your county club, via reports and avifaunas, is a great way to get an understanding of what's 'expected' and what isn't.

New ground

Finally, it's worth considering exploring new ground in June. Given that this is one of the quieter months of the year, it can be a good time to go 'off-patch' and explore parts of your local area that you've not visited before, with little risk of missing something good on your familiar routes. And you never know, you may just bump into a patch goodie or rarer breeding species in a previously unknown area... Any breeding records from 'new ground' sites are of tremendous value too, especially in parts of the countryside that have no or very few previous ornithological records.

Making the most of mammals

If you're reading this book, then of course you're a keen birder. But many of us are all-round naturalists who enjoy studying and observing different forms of wildlife. Midsummer represents a great time to focus more time on such groups – butterflies and moths are at their most obvious and diverse (see page 119 for more on this). However, it's also a great time for seeing wild mammals.

Mammalian diversity is not exactly great in Britain, but the long days of summer and often fine weather makes June a good time to embrace the twilight and go looking for them (dusk is generally deemed the optimum time to go mammal watching, but it does vary depending on the species). There are different ways to watch mammals, from spotlighting and camera traps to bat detectors and small-mammal traps.

Midsummer is the best time of year to see Stoats and Weasels, for example, which will be feeding young. Fox cubs might be observable and staking out a Badger sett might be rewarding. Several species are particularly active during the summer, too, including Hedgehog and both dormouse species (Edible and Hazel). If you live along the coast, the summer is typically the peak period for cetaceans – you may even be lucky enough to have a headland on your patch from where you are more likely to see passing marine life. Common Seals breed during June, too.

Mammal watching can be rewarding in the summer when breeding activity is high. It's a good time to look for Stoat (left) and Roe Deer (right).

June is perhaps the best month of the year for bats. Seventeen species regularly breed in Britain and, in good midsummer weather, they can seemingly be everywhere at dusk, from gardens and parks to waterbodies and woodland. Going out with a bat detector – of which a wide range in terms of quality and cost, exist – is a great way to enjoy and appreciate these creatures, as is joining your local bat group. In time you'll learn the different intricacies of various species. For example, Noctules are often on the wing while it's still light but Daubenton's Bat is truly nocturnal, while pipistrelles are roughly in between. Mammals offer a great alternative for summer wildlife watching if patch birding is quiet.

07 July

Occasionally referred to as the start of 'birders' autumn', July is not a complete midsummer break for the patch-watcher. Post-breeding dispersal and return passage is underway now, and gradually builds during the month to make it a rather exciting time to be in the field. With the increasing dewy mornings, slow fading of flowers and grasses, and marked reduction in birdsong – with some species like Cuckoo and Nightingale already having departed breeding grounds – it can feel like a significantly transitional month, even if non-birders tend to be baffled at the idea of it being 'autumn' in July!

If you've used June to unwind a bit from patch duties, July is a good time to slowly ease yourself back into the routine of checking your patches more thoroughly. There's plenty to go looking for, whether you're inland and or by the coast. The slow and enjoyable journey towards autumn 'proper' begins here, so make sure to savour it from the very start.

Returning from the north

As touched on in the June section, return wader passage is already underway and builds during July. By the end of the month, the autumn's movement of shorebirds is already beginning to peak, at least for certain species. Whatever the weather, it's a good time to check waterbodies. It's a particularly good month for Green Sandpiper – July might represent the peak time of year to score this species on one of your less-visited patches, even in dry locations, something entirely possible given the unfussy nature of this underrated species.

Another species on the move at this time is Black-tailed Godwit, as birds return en masse from Iceland. North-westerly winds and cloud cover or precipitation in particular can be productive for dropping in waders moving back from Iceland (not just godwits). Keep an eye out for juvenile Black-tailed Godwits of the subspecies *limosa*, which breed in England in low numbers as well as in mainland Europe. They disperse from breeding grounds earlier than Icelandic breeders, so if you see a juvenile in July, it's highly likely to be this form.

July sees Black-tailed Godwits return from Icelandic breeding grounds.

Most waders in July will be failed breeding adults, though some British-born juveniles of certain species may be in evidence, particularly during the end of the month. Among those to keep an eye out for in July are Whimbrel (which can be picked up migrating over almost anywhere – I've had flocks over woodland before!), Greenshank, Little Ringed Plover, Dunlin, Redshank and

Common and Wood Sandpipers. If it's been a poor breeding season in Iceland, northern Britain or Scandinavia, certain species may be present in greater numbers earlier on.

If you're a landlocked birder, July can represent the best month for certain species. It's by far and away the best month of the year for Redshank on my patches – an otherwise rare visitor to my area – and is particularly good for Oystercatcher, Whimbrel and Greenshank as well. If you're on the coast or are lucky enough to have a decent wetland in your area, then much rarer prizes should be considered, including North American species or those of Eurasian origin such as Terek Sandpiper.

On the subject of wetlands, it's worth mentioning Common Scoter. Drakes especially will move back to non-breeding grounds in the Irish Sea and English Channel in July and, if there is overnight inclement weather, such as fog or rain, it's worth checking inland waterbodies for grounded individuals, much as is the case for this species in March and early April.

Early vis-mig

Although the peak vis-mig months of September and October are a little while off, there are a few earlier autumn niches that the migration connoisseur can enjoy in July, sometimes providing entertainment during an otherwise slower month for movement of smaller birds.

Perhaps the most exciting vis-mig at this time is the movements of Swifts. This can be detected across Britain, though it is more pronounced along the east coast. The best days for passage are usually during the last week or so of June and the first half of July, though big days can occur up to the end of the month. It's very much weather dependent. Ideally, a fresh south-westerly or westerly wind, preferably with precipitation or low cloud cover of some form, will be in play. The action can be congested or run all day – try to react to the weather and watch from a high point in your area or patch. I've enjoyed some brilliant Swift watches in July, including movements of several hundred birds – not bad for Surrey. I've been treated to some skywatching surprises while watching such big days of passage, too, including flyover Curlew and Black-tailed Godwit far from any waterbodies.

Large movements of Swifts can occasionally be enjoyed during July.

Crossbills may disperse widely as early as July when the cone crop in their breeding region fails.

That so many Swifts move around during the height of the breeding season begs many questions. Are these locally breeding birds just moving around in front of depressions as they hunt for flying prey, or overshoots making their way back to the continent, or perhaps non-breeding birds making an early start south back to their wintering grounds? Ringing has confirmed that most are first-summer birds, so maybe roving or early returning non-breeders is the answer. Whatever the case, it's a classic example of bigger-picture migration that can be enjoyed on your doorstep, with luck.

Generally speaking, vis-mig diversity is low in July, but a couple of other species are worth a mention: Siskin and Crossbill. In irruptive years for these species, movements can begin in July. In 2015, which proved an extraordinary autumn and winter for Siskin, I first noticed vis-migging individuals and small flocks as early as July.

The same is true of Crossbill, which is a more nomadic and irruptive species. In good years, July can be a fantastic time to connect with it – and if you live in an area where the species is scarce, then a vis-mig session might be your best chance of scoring. Even where I live, in heavily wooded Surrey, Crossbills can sometimes be difficult to connect with earlier in the year – and more than once I have sought them out in July and managed to locate birds. Similarly, I have had flyover records at sites that you wouldn't associate with the species in July, often on fine, warm days.

First bush-bashing

Towards the end of July, passerines will start moving around, at least locally and regionally – July isn't too early to encounter a Redstart, Pied Flycatcher or Wood Warbler working its way back south, for example. From mid-month, lemon-yellow young Willow Warblers may be encountered – really smart-looking birds and always an encouraging sign of the passerine pack being shuffled. By the final few days of the month, many post-breeding dispersing birds are at large, including species that might not breed anywhere near you. Grasshopper Warbler is a good example near me – I've had a few late July records of this Surrey scarcity.

It's therefore worthwhile to get into the mindset of bush-bashing – quietly and patiently working areas of hedgerow, scrub and woodland edge to try and detect an out-of-place species. Sometimes a simple bit of patch gold will do, maybe a Sedge Warbler far from

By late July the first signs of passerine migration can be witnessed, with warblers, such as Grasshopper Warbler, beginning to disperse from breeding grounds.

water in some arable farmland, but scarcer goodies are not impossible at this time of year. The first gatherings of mixed flocks are in evidence by now as well, something that will be touched on further in the August section.

Other birds, such as finches, will also begin to gather in flocks. Sifting through them now is good practice ahead of the autumn proper, when bigger prizes will be in play. Indeed, July can be a fun period to dedicate some time to counting flocks. Many birds gather in large post-breeding numbers, especially Woodpigeon, corvids and Starling. You may be able to set some personal high count records at this time of year, some of which might be of county-level significance.

Get among the gulls

July is a good month for gulls. Yellow-legged and Caspian Gulls on the continent fledge earlier than our breeding large gulls, so early July is an ideal time to look out for juveniles as they begin to disperse. In the south and east of England, any juvenile gull in the first half of the month is worth perusing as it may be a Yellow-legged; Caspians tend to turn up a little later. Check loafing flocks of large gulls, which may gather on a reservoir, estuary, farmland or playing fields. Throwing bread can be useful to entice birds closer, too. For many of us, either of these two species represent a pleasing find – be sure to brush up on the ID first. Getting good photos can be

Looking for fresh juvenile Yellow-legged Gulls is a fun activity for southern-based patch birders in July.

handy if you're less experienced with immature gulls.

Smaller gulls like Mediterranean, Black-headed and Common are also dispersing now, and high counts can be recorded away from breeding colonies – it's a good opportunity to collect Darvic ring data which provides valuable information on bird movements. And, of course, it can pay to sift through these larger-than-usual gatherings for rare nuggets of gold – a fellow Surrey patch-watcher found a mid-July Franklin's Gull amid a large flock of Black-headed Gulls, a first for the county no less!

It's also a good time for terns. As with waders, failed or non-breeding adults will be moving back south and, depending on where you live, it can be a good month to score inland goodies such as Little or Sandwich Terns. July is also a good month for rarer species, including Roseate and, in particular, Caspian Tern, a national rarity that can turn up almost anywhere and for which records peak in July. It may even be that Common Tern is a goodie in your area – many juveniles will be on the wing by late July, and you could pick up a wanderer at this time of the year.

July can sometimes turn up left-field rarities such as a Caspian Tern.

Insect interest

July, particularly the quieter first half of the month, is maybe the best time of the year to engage with insects. Broadly speaking it's the best month of the year for butterflies, for example, certainly in terms of species diversity and abundance. I turn my attention to butterflies every summer and I make the effort annually to see a few favoured species. If you're new to butterflies, July is a great time of year to get in the field and learn.

It's also a brilliant time for moths – and moth trapping. July offers a greater abundance and diversity of species on the wing than any other month. Moth trapping can seem a daunting commitment to a beginner, but if it's something you've been curious about it's worth experimenting and giving it a go. Moth traps come in all shapes and sizes, and are easily available to order online. Perhaps the most familiar design is the Skinner trap, which consists of a simple box, with glass or clear Perspex sloped lids and a gap in the middle beneath the light bulb fitting. They are also easy to build yourself.

Moths are attracted to the light and then funnelled down into the box, where a bundle of egg cartons offers somewhere for them to spend the night, so they're ready for you to uncover in the morning before releasing them. You'll need to be up early to work through the moths you've caught while they're cool and calm – but being up early is nothing new to the patch birder!

Once your moth trap is fired up you will potentially attract all sorts of species from across a wide area, including passing migrants, but you will certainly see a bias towards the species that live on the particular trees and plants in the immediate vicinity. If you want to catch Oak Beauty, December Moth or Merveille du Jour, for example, you need Oak nearby. If you have willowherbs or fuchsias in or near your garden, you may be lucky to catch the gorgeous Elephant Hawk-moth in July.

Depending how far down the insect rabbit hole you'd like to go, July is the peak time for activity of other groups, with everything from ladybirds and hoverflies to dragonflies and bees all busy hunting, or pollinating and generally contributing to the thriving web of life on your doorstep.

It's also a great time to get familiar with wildflowers, such as orchids. Many flowers begin to look tired by August (which is also when patch birding is in peak mode!), so the quieter days of July can be a good time to enjoy them.

As with birds, be sure to record all your insect and wildflower observations. In Britain, iNaturalist and iRecord (see page 172) are particularly popular and useful apps for such taxa. All the data will help county and national recorders build up a better picture of the populations in your area.

Left: Dark Green Fritillary **Above:** Elephant Hawk-moth.

Pied Flycatcher is a classic August species that can turn up just about anywhere.

08 August

If you asked a group of patch birders what their favourite month of the year is, August would be a likely reply from many. Post-breeding dispersal and return migration gather increasing momentum as the month rolls on, with species diversity – and the possibility of the unexpected – high. On top of this, the days are still long and, often, warm and pleasant, making August an ideal time to be in the field. Misty, dewy mornings with the first whiffs of ripening fruit in the air offer signs of things to come, but often the weather is settled, with insects still on the wing in decent numbers. It's a wonderful combination of seasonal senses.

Autumn is a gradual process and, even though the early signs can be detected in July and even late June, things crank up significantly during August. By the end of the month, a huge variety of species are in play and birding days can be dynamic and exciting. I always find the August Bank Holiday weekend to be a particularly productive one, for example. Head out to the patch as much as you can this month!

Mixed flocks

Post-breeding dispersal and birds moving through the countryside in August mean that some species, whether regionally scarce or merely locally surprising, can appear anywhere. Many migrants latch on to roving mixed-species flocks, themselves symptomatic of this time of year. These varied squadrons trigger intrigue and anticipation in the patch birder. Normally a lemon-yellow juvenile Willow Warbler or two will be dug out, but why not a Firecrest, Pied Flycatcher or even something rarer? Particularly inland, such gatherings can prove to be treasure-troves for the eager patch watcher of species that join roving flocks is long. Warblers are classic 'sentinel' species, with *Phylloscopus* the genus most typically encountered, although Blackcaps often tag along, and I've seen Common Whitethroat and Garden Warbler in such groups too. Young birds especially will readily join mixed flocks – indeed most gatherings you find at this time of year will hold a juvenile Chiffchaff or two. There is a

Fresh 'lemon-yellow' juvenile Willow Warblers start appearing in late July and early August.

Careful scrutiny of mixed flocks may reveal an energetic Firecrest, a striking bird that is still notable in many areas.

degree of 'pick and mix' when it comes to these flocks, as you're never quite sure what might be tagging along with a group – and sometimes there is treat or two to be found!

Scarcer options include Wood Warbler – for those who live far away from their diminishing breeding range, scouring mixed flocks offers maybe the best hope for unearthing this pretty warbler locally. Although far rarer, Barred and Greenish Warblers have been found in or, more likely, on the periphery of mixed flocks at this time of year; indeed, no fewer than 176 Greenish Warblers were logged in Britain in the month of August prior to the species' demotion from the ranks of BBRC rarities in 2005. Icterine and Melodious Warblers are not impossible either – dream big and aim low is my birding mantra!

Goldcrest and Firecrest are usually eager participants in mixed flocks and, of course, the latter is still a sweet find in many areas. September is also good for this species as birds from their ever-growing British range disperse.

Both Pied and Spotted Flycatchers are in play should you chance upon a mixed flock. Usually rather solitary birds, multiple tag-alongs can sometimes be found in these congregations. In 2020, one lucky birder counted a minimum of 20 Spotted Flycatchers among a mixed flock in my home county of Surrey – this same flock also held two Pied Flycatchers (as well as 120 Blue Tits, 80 Coal Tits and 65 Chiffchaffs!).

It's not just migrants that can be of interest in mixed-species flocks either. Locally uncommon birds are very plausible – Lesser Spotted Woodpecker is a prime example, and I've seen plenty in large gatherings, often seemingly far from suitable habitat or a known site. Marsh and Willow Tits, both sedentary species, are also realistic, especially earlier in the season when juveniles disperse.

So where do you find mixed flocks? The answer is quite simple: anywhere. There is no specific habitat type, though a few factors are involved. The bigger the flock, the better the chance of something unusual, so focusing on areas where tits are abundant is a good start. Wooded landscapes are obvious places to begin your search, though any well-vegetated area – be it scrubland, a series of hedgerows or even a reedbed – is likely to be a good spot.

The flocks have formed to find prey together, so seeking out areas with a good amount of insect activity is worthwhile. This means spots near water in particular – even a small pool in the wider landscape can make a big difference in terms of attracting feeding flocks. Try also to note areas on your patch that are prone to catching the sun. These suntraps can be a magnet for insects, and it doesn't take long for birds to locate them. In prolonged hot weather, when smaller waterbodies dry out, birds may concentrate at the edges of the pools that remain, to bathe and drink.

It's important to maximise your own ability to observe things. So, while plenty of roving flocks may be located in woodland (for example), being under the canopy is not helpful when it comes to keeping tabs on a fast-moving mass of birds. I find the best areas are often marginal ones, or at a crossroads of habitats – a mixture of tall and low vegetation, relative openness, plenty of cover, water of some sort and a healthy supply of sunshine are perfect.

Topography can be used to your advantage as well, especially when seeking out more unusual species.

Woodland edge, with a mix of bushes, thickets, hedgerows and mature trees, are excellent places to look for mixed flocks in late summer and early autumn, especially on sunny days.

Locating a mixed flock on high ground is more likely to yield a Pied Flycatcher, for example, as migrant birds hit these 'vertical inland headlands' with more frequency. River valleys are also well worth exploring.

There tends to be two main peaks of feeding activity among feeding flocks at this time of year: in the morning and again in the early evening. The morning period is often the most fruitful, with the possibility of digging out newly arrived migrants that have joined a local flock. You don't need to start looking too early, though – ideally let the day warm up a bit so insects are active. From at least two hours after sunrise until mid/late morning is generally optimum. A second peak is often detected in the early evening, especially involving birds feeding up before moving on overnight. A still, dry and mild day is the most suitable for finding feeding flocks.

Weather alert

August heralds the beginning of the main autumn migration season, which can last all the way to the start of December. As the month goes on, migration starts to pick up and the variety of species arriving in Britain increases dramatically. The breeding season is over in northern and eastern Europe, and most of the birds on the move are youngsters on their maiden migration flight. First-time migrants are especially likely to lose their way if blown a little off-course by moderate winds, and then turn up outside their species' usual range. So it's highly recommended to stay tuned into the weather forecasts, both at home and on the continent, from now all the way through the autumn, to identify the best times to seek out 'drift migrants'. As reiterated in the weather section, playing the conditions to your advantage is a key part of patch birding – and it's arguably never as handy as in the autumn.

Easterly winds may produce drift migrants from Europe, like the scarce Ortolan Bunting.

In August, most birders will be hoping for easterlies. The biggest arrivals are usually later in the month, but good conditions earlier can produce bumper numbers of classic patch fare such as Wood Sandpiper and Pied Flycatcher. Keep an eye out for a high-pressure anti-cyclone developing over northern Europe, which usually signals clear skies and good winds for a flight south over Denmark and along the North Sea coast, down through the Netherlands. If these southward-moving birds then meet a low-pressure system from the west it can result in a large fall of migrants in Britain. Typically, they will hit the Northern Isles or east coast first, but it doesn't take long for birds to filter inland.

Typical August arrivals include many species mentioned in the mixed flocks section, such as Redstart and Pied Flycatcher. However, August also represents the first time in the autumn that real goodies are in play – think Wryneck, Red-backed Shrike, Bluethroat and Ortolan Bunting, as well as some of the warbler species covered already. And it's not just wind conditions for drift migrants that matter – rain and cloud is also important, as we'll consider next.

Waterbodies

It pays to never be too far away from a waterbody during August. At this time of year water can provide a wide variety of migrants, from waders, terns and wildfowl to raptors, gulls and crakes. Good birds can be found on fine late-summer days – but keeping an eye out for wet or cloudy mornings is particularly useful, as these are the conditions that are most likely to drop migrants.

By August, a wide variety of waders will be on the move, with both adults and, increasingly, juveniles present. Common and Green Sandpipers are ubiquitous at this time, with Greenshank and Black-tailed Godwit also passing through in numbers, along with Dunlin and, later in the month, Curlew Sandpiper and Little Stint. Wood Sandpipers can peak in August, too – they will occasionally be unfussy with where they pitch up, so keep checking any pools or the edges of any waterbodies.

The wind direction may dictate things a little: winds from the north-west may produce birds from Iceland and Greenland, such as Black-tailed Godwit and Whimbrel, while easterlies are better for species like Wood Sandpiper, Ruff and Little Stint. If you have a

Wader action is firmly underway during August, with Ruff being a species that occurs at this time.

Migrant terns, like this Sandwich Tern, can be cause for excitement at this time of the year.

particularly good wader site in your area then August can be a fantastic month for variety – and surprises.

August is probably the best autumn month for terns. It's a classic time for Black Tern, for example, especially in easterly winds which, if strong and persistent enough, may even send a few White-winged Black Terns into the country – a national scarcity gettable both for the inland and coastal patch birder.

Sandwich Terns disperse around the coasts after fledging, some as early as late June, and numbers at some staging areas peak by mid-August. Although a coastal bird, it's a real possibility inland, even if it's a species that never tends to hang around for long. You're likely to hear one before you see it – that's been the case for me with virtually all of my patch 'sarnies' – so bear that in mind.

Arctic Tern, a desirable patch bird for many between late April and early May, leaves its breeding colonies from mid-July to mid-August. It is not so associated with autumn migration – not least because many head due west from east-coast colonies to reach the Atlantic – but is worth considering if you have a build-up of Common Terns on your patch.

Most Little Terns leave their breeding areas soon after chicks have fledged and have departed Britain by the end of August – this species is a possibility in July. However, it remains in play during August, along with a real patch goodie – Roseate Tern. Although very rare inland and barely much more regular along the coast away from regular haunts, any large gathering of terns in August should be checked for it. Far rarer dream patch finds to consider include Caspian, Gull-billed and Whiskered Terns.

Bird of prey bonanza

Raptors are on the move in August. Aside from Hobbies and Honey Buzzards, which are still finishing up, most species are done with breeding and populations are swelled by fledged youngsters. The often fine weather means it's a good time to embrace the thermals and rove around. Young Goshawks are prone to exploring beyond their natal areas in early autumn and may be found over farmland or near waterbodies – keep an eye out for an explosion of alarmed corvids or Woodpigeons. Young Peregrines might be evident, too.

August is a brilliant month to see Osprey on your patch. Wherever you may be in Britain, Osprey is a realistic target – migrating birds may be chanced upon anywhere. If available, choosing a suitable vantage point for skywatching is recommended to give a good all-

August is a great time to connect with Ospreys as they migrate south.

round view and increase your chances of intercepting this exciting raptor (or another migrant bird of prey).

Although raptors are typically associated with nice, fine weather, it is often more inclement conditions – like a strong headwind or low cloud cover – that produce more visible Osprey passage. Autumn migration tends to be a more leisurely affair, too, and birds will happily stage at multiple sites on their way south, sometimes lingering for several days.

The aforementioned Honey Buzzard is a classic late August bird. Adults will be well on their way south by this time, and for many patch birders this species represents a major prize. Juveniles will be migrating well into September. Rare harriers are not out of the question, either, with the now desperately rare Montagu's Harrier a possibility, along with Pallid Harrier, the latter remaining on the cards well into October. Bear in mind that raptors are best sought out from mid-morning until mid-afternoon.

A real prize in August and September for a patch birder virtually anywhere in Britain and Ireland is Honey Buzzard. Juveniles like this bird migrate south later than adults and can sometimes be seen as late as October.

Eyes to the sea

Seawatching offers a whole new dimension for the patch birder, complete with different seasonal- and weather-based focuses. Whether you're situated in western or eastern areas, early autumn – August and September – are excellent months for seawatching.

A fantastic combination of species can be seen. In western areas, local birds dispersing from nearby colonies make up the bulk of numbers, but with Arctic-breeding species (such as Grey Phalarope and Arctic Tern) and southern hemisphere wanderers (like Sooty and Great Shearwaters, Wilson's Storm Petrel) mixed in, as well as some Mediterranean and Macronesian species (Balearic and Cory's Shearwaters). In eastern locales, waders migrating south from the Arctic, the first trickle of wildfowl arriving from the Baltic and skuas leaving their Scandinavian and Russian breeding grounds make for a veritable smorgasbord of activity.

Generally, morning and late afternoon into evening are the best times of day to seawatch. The trick is to look for suitable onshore winds, which will bring seabirds close and into view. Frequent bands of rain help with this, as long as it isn't too heavy or excessively hindering visibility.

Wind direction is of major importance and varies depending on where you're watching. The right equipment is essential, too, especially if you're undertaking a lengthy stakeout. Make sure you have warm, waterproof clothing, a telescope and something comfortable to sit on, like a chair or cushion.

If you happen to live by the sea, seawatching is an important element of late summer birding, with goodies such as Balearic Shearwater (left) and Wilson's Storm Petrel (right) possible in the south-west.

09 September

After the gradual build-up of autumn excitement during the last couple of months, September is the month when migration is in full swing – and the shift in the season is undeniable. Birds like hirundines and warblers will be prevalent by the start of the month; by the end, winter thrushes, finches and wildfowl will be arriving. It is a month of diversity and, for many, the most dynamic time of the year, as species composition and the sheer volume of birds reach their peak. Days on patch at the start of September can feel very different to those towards the end of the month, as we say goodbye to the British summer.

At times, September – and the vast array of potential it presents to the local birder – can almost seem overwhelming, such is the variety of field methods you can deploy at this time and the number of possible target species. Meanwhile, as the month goes on, the bird news pages will be increasingly filled with reports of incredible rarities from far-flung places across Britain, as well as goodies closer to home, sometimes making you unsure which direction to take. In short, it's a busy time for British birding – and a month that makes it worth getting on patch as much as you can!

The vis-mig season is underway by September, with species like Grey Wagtail detectable overhead.

Vis-mig begins

Serious vis-mig commences in September, especially the second half of the month. This form of birding takes patience and skill, but is incredibly rewarding, especially for the local birder. Species verging on unthinkable at other times of the year are on the cards, as mass southbound migration takes place. What conditions are best will vary depending on where you live, but generally some cloud cover is useful. Nailing the identification of migratory species in flight is a challenge, especially given most birds passing through can be little more than a brief flicker of feathers. However, it is a thoroughly relaxing pastime, often undertaken as the day begins and allowing the birds to come to you. Some days can be brilliant, others might be quiet, but the mere possibility of a big day of movement or a rare species makes the stakeout worthwhile over time.

Among the earlier peak vis-mig season movers are pipits and wagtails. Early September is a great time to connect with Tree Pipit, for example, and if this is a species not found in your local area then it's well worth staking out your favoured vis-mig spot. The flight call is distinctive. Meadow Pipits also move during September, with numbers peaking in the middle of the month.

September is a good month for encountering raptors, including Hen Harrier.

Sometimes, passage can be enormous – even inland – and produce quite a spectacle. Yellow Wagtail is another species symbolic of September vis-mig, and it's a sign of wider migration activity if you hear one flying overhead at this time of year. It is possible to detect subtler migrants, too, including Grey Wagtail, which is a notable mover during September in my local area and a bird you may not necessarily associate with migration.

September is the best month for large hirundine movements, too. Typically, these take place in coastal areas, where counts can be spectacular, offering a wonderful wildlife display. A bit of a breeze often helps, especially a headwind. If a big hirundine day is in play, it's worth investing time at your best vis-mig spot – you may be rewarded with a remarkable count.

Towards the end of September, thrushes and finches begin to arrive. However, they are more strongly associated with October and so are covered in full detail in that month's account. A wide variety of species is possible in September, though, including rarities. As a result, it's worth keeping your wits about you while vis-migging, including having your camera nearby and ready to go, and maybe even a sound recorder running while you watch.

Raptors are well worth targeting in September, too, though generally they move later in the day than during a typical (morning) vis-mig session. We covered them in the August section, but arguably there is more raptor diversity in September. As well as southbound Hobbies, Ospreys and Honey Buzzards, more typically northern species begin to move at this time, including Merlin and Hen Harrier. The latter will disperse from breeding grounds in northern Britain and Europe and filter south through the country during September

Wader numbers will build at coastal sites during September.

and October. Considering the chance of the much rarer Montagu's and Pallid Harriers (and even Northern Harrier from North America), be sure you run through the identification features of any harrier you see.

The golden rule of patch watching, context, is also important here, and more regular species like Marsh Harrier may represent a good local bird for you. Indeed, if there's a raptor species missing from your year list, September's a great time to try and get some sky-watching in during the middle of the day, or stake out favourable areas at dusk ahead of birds coming into roost.

Wader specials

Waders have featured in an awful lot of these monthly sections, but there's good reason for it! Not only are they wonderful birds, many of which represent exciting and quality patch prizes, but they are also in play at all sorts of different times. However, September is perhaps the best month of all for encountering a range of species. At this point, juveniles are passing through the country or dispersing from breeding grounds, swelling numbers.

Many of the species listed in August and July are still realistic possibilities during September. As well as these, the availability of rarer species peaks at this time. Three species that breed in Scandinavia and the high Arctic – Spotted Redshank, Curlew Sandpiper and Little Stint – peak in Britain during this month. All three are notable at any wetland, coastal or inland, especially the two last-named species.

Meanwhile, from North America, September is the best month of the year for Pectoral Sandpiper – a Nearctic species that can occur just about anywhere. Other, rarer North American (and Siberian Arctic) waders are possible at this time – no fewer than 23 Nearctic shorebird species have been recorded in Britain, and September is the time to dream big when it comes to patch-waders!

As ever, inclement conditions are handy for grounding waders, especially inland, though regular checking of your best shorebird areas in any conditions are advised. If you're by the coast, then multiple checks a day of prime wader hotspots are worth undertaking, while keeping an eye on the tides. Waders come and go on passage and can sometimes drop in for only short periods.

Carefully scan big flocks of commoner species, which may contain rarer birds. Inland, even less-expected waterbodies can yield shorebird results when conditions are optimum, so keep an open mind. Searching through gatherings of waders is perhaps never more rewarding than in September. For starters, observing and learning the differences in adult and juvenile plumages can be undertaken, and the chance of picking out something scarcer is at its highest.

Rare waders, such as Pectoral Sandpiper, are worth keeping in mind during the peak weeks of autumn.

Autumn storms may result in species like Sabine's Gull turning up at coastal locations or even inland.

And, of course, so many other wetland goodies are in play in September, not just waders. It's a great month for Spotted Crake, for example, while rare and scarce herons are on the table during this time, not to mention other possibilities like Garganey, Black-necked Grebe, Glossy Ibis and Black Tern, to name but a few. In short, make sure you're checking your waterbodies as often as you can during this month.

Inland seabirds

Another scarce wader that peaks in September in Britain is Grey Phalarope. This delicate but seafaring wader is particularly associated with stormy conditions, when Atlantic gales can displace birds along the coast and well inland. In the immediate days after such a storm, it's a species to bear in mind. It's a rare prize for any patch-watcher, but also a realistic one if the weather aligns – and especially if you're on the coast or watch a decent waterbody. If a day or two of strong westerly gales has abated it might be worth hitting your local reservoir, lagoon or estuary. Young birds can often be extraordinarily approachable, often making for truly memorable encounters. It is one of various coastal birds that become possibilities for the inland patch birder during September.

Indeed, September is one of the best months for various pelagic species, including skuas, Leach's Storm Petrel, Manx Shearwater and Gannet. Of course, all of these are very rare inland and are not to be expected – but they are worth considering. Scarcer and more pelagic gulls, including Little Gull, Kittiwake and even Sabine's Gull, are possibilities as well.

A storm in the North Sea or to the west of Britain can result in seabirds being blown inland. Check reservoirs and lakes – the bigger bodies of waters – during and immediately after these bad spells of weather. By carefully sifting through the Cormorants, you might pick out a Shag, or a Kittiwake or Little Gull might be flying around with the Black-headed Gulls. Being alert for seabirds during vis-mig sessions is important, too. Skuas in particular are known for occasionally migrating overland, and exceptional species like Cory's Shearwater have even been seen on inland watches before. Keep an eye on the bird, news, too. If there has been a push of skua passage on the east coast, for example, birds might cut overland in the following days.

If you patch-watch on the coast, then it goes without saying that your chances of an encounter with species such as Leach's Storm Petrel, Sabine's Gull, Grey Phalarope or a rarer skua are greatly increased in and around periods of stormy weather. Autumn seawatching in rough weather is not for the faint-hearted – but it can be immensely rewarding. Indeed, September can be a highly dynamic month for seawatching, whether the conditions are rough or not.

September scarcities

In the August section, we discussed checking through mixed-species foraging flocks, and later in this section we cover scanning fence-lines and hedgerows. Both these methods are as relevant in September as they are in August – and, if anything, there is more reason to deploy them at this time, as a suite of rare and iconic patch possibilities are in play. 'Bush-bashing' is worth plenty of time this month.

Among the more expected birds found during such field sessions include warblers, many of which are moving south at this time and can result in species in atypical habitat (like *Acrocephalus* warblers in farmland), localised species outside of their normal range (like Dartford or Wood Warblers) or even scarcities. Patience is required while working through warbler flocks. Flycatchers are also in the mix, often on the periphery of mixed flocks.

However, there are rarer goodies that peak in Septemberso this is perhaps *the* month to seek them out on your patch. One classic September bird that's possible almost anywhere is Wryneck. Passage Wrynecks can turn up in a variety of habitats. All that's required is food, chiefly ants. Open areas are often favoured as a result, be it grassland, arable farmland, headlands, islands, beaches and even gardens with large lawns. Working sites slowly and scanning open areas of ground is a good method. I've found Wrynecks by inadvertently disturbing them from feeding on the ground on a couple of occasions. Wrynecks do seem to turn up at the same sites repeatedly and that can be worth considering when looking for one. Perhaps a little research into where birds have historically shown up in your local area is worth a go.

September is the best month to try to find a patch Wryneck.

Red-backed Shrike is another species that, like Wryneck, is firmly on the cards during September and can turn up anywhere. Areas of scrub and dense hedgerows are optimum, but passage birds are not fussy and may turn up in rather unassuming locations. Scan hedgerows and fence-lines – a method that can also produce Whinchat, Wheatear and Black Redstart.

For species like Wryneck and Red-backed Shrike, it can pay to keep an eye on the weather (as detailed in the August section). At this peak time of year, easterly or north-easterly winds associated with high pressure over Scandinavia can produce big arrivals of passerine migrants in Britain from Fennoscandia– including scarcities such as Pied Flycatcher and Icterine Warbler. These are the perfect time to go looking for patch rarities. Keeping up-to-date with local and national bird news pays, too, as it can paint a picture of what is around and what should be on your radar. Indeed, by being switched on and

By the end of September, most Whinchats have left the country.

with perseverance, even rarer prizes are possible at this time of year.

Farms and fences

Fences aren't always the most obvious places to look for birds, but in September (and August) they are worth keeping an eye on. Fenceposts and fence-lines are often used as perches by migrant passerines, and classic patch fare like Wheatear, Whinchat and Black Redstart can all be found by scanning with binoculars along such structures.

At this time of the year many of these species, whether adults having completed their post-breeding moult or fresh juveniles, nicely reflect the colours of autumnal vegetation, often being a mix of soft sandy, orange or pinkish tones. Many individuals might be approachable, too, and it marks a great opportunity to take time and enjoy views.

Many of the most productive 'fence-line sites' will be on farmland, and it's worth paying attention to livestock in September. Yellow Wagtail is a classic species of this time of year, and sometimes large flocks can be found around cattle herds. An increasing vagrant to Britain is Eastern Yellow Wagtail, so if you encounter a particularly monochrome bird in a flock then scrutinise it carefully (and ideally obtain sound-recordings). Citrine Wagtail is another possible September rarit, though perhaps more likely at a wetland. It's worth working through Pied Wagtail flocks, too. Identifying autumn White Wagtails is very tricky, but is do-able and, in some areas of Britain, they are doubtless under-recorded at this time of year. The best feature to look for is the paler grey rump.

10 October

Autumn is still in full flow and it's all change in the birding world. In a month that will see the last of many summer migrants depart our shores as well as the first arrivals of winter visitors, there is so much for the local birder to get their teeth sunk into. Turnover can be daily, with each day providing a different migration element to the next. Even though the days are shortening rapidly, there is still enough daylight to allow for lots of time on patch.

For many birders, October is *the* best month of the year, offering dynamism and excitement as return migration continues. Furthermore, it is perhaps the likeliest period of the year for encountering rarities, be they national megas or local treats. On top of all that, it's probably the most enjoyable period for vis-mig – that excellent birding niche that can be experienced almost anywhere, and is thus prime for patch-watchers.

October is the month of peak vis-mig, when flocks of migrant birds, like these Redwings, can be seen overhead.

Peak vis-mig

October is the peak month for vis-mig in Britain. It is when the biggest arrivals of winter thrushes and finches tend to occur, but also sees the last summer migrants head south. On top of that, wildfowl are migrating, raptors are dispersing and many more are on the move. If you can dedicate some time to vis-migging in October then definitely do – it'll perhaps be the most rewarding form of local birding you experience this month. Optimum ways to do so are covered elsewhere in this book, including pages 130–132.

Watching the skies on an October day of busy migration is akin to panning for gold, sifting through commoner species – themselves often providing quite a spectacle – for something rarer. October is usually the main month for thrushes arriving from Scandinavia (though November can also be good). Redwing is typically the most numerous species and, on big days, movements provide a wonderful spectacle, with birds' *sseep* calls filling the air. Redwing, which is synonymous with autumn vis-mig, moves south-west out of its breeding grounds in Fennoscandia, down through the Netherlands into France and Iberia, and into Britain. The largest movements are usually seen on the east side of the country, with birds often arriving first in the far north. However, in 2021 a record-breaking count of migrating Redwings in Britain was made in Surrey, with an astonishing 34,727 logged at Leith Hill on 13 October just beating the previous British record set at The Pinnacle, Sandy, Bedfordshire, on the same date in 2009. These two counts show that you don't need to be on the coast to enjoy migration spectacles.

Fieldfares tend to move in smaller numbers (and they arrive a little later than Redwings), though big days can still be experienced, especially in eastern areas. Although not always recognised as winter thrushes, both Blackbird and Song Thrush also migrate from northern climes to Britain to swell our winter population. These migrants are often more discreet than *sseep*-ing Redwings and *chak*-ing Fieldfares, but sometimes impressive numbers can be detected moving overhead.

Large migrating flocks of these winter thrushes can sometimes act as carriers for other rarer species – and it is always worth checking through feeding parties. A favourite patch prize is Ring Ouzel. For many patch-watchers, October is the best month to encounter this species. They may be flying over on their own, in small groups or in with flocks of Redwings and Fieldfares. Persistent scanning is important. Numbers tend to vary year on year; in good years, with a healthy berry crop, it's worth checking your local rowans, hollies and junipers, which may yield feeding birds.

Far greater prizes are not impossible, although they're more likely in the Northern Isles or along the east coast. Examples include Eyebrowed Thrush – one was found feeding with Redwings in Highland in late October 2021. October is also the best month for both Black-throated and White's Thrush records in Britain – but these of course cannot be expected, even though the former species has a knack for appearing inland.

Many finch species move en masse during October. It is a great month for

Opposite page: Finches arrive en masse in October, including Bramblings from Scandinavia.

catching up with Brambling, Hawfinch and Crossbill, for example. Knowing the flight calls of these species is extremely useful. Many finches are prone to influxes linked to food sources both in Britain and on the continent. In some winters, it may seem like numbers of certain species are fairly low, then higher than usual the next year. These variances can often be detected during the autumn vis-mig season – if you've encountered lots of flyover Bramblings, for example, then perhaps it'll be a good winter to check your local beech woodlands. Hawfinch is a species that, if it doesn't occur in your local area, is highly desirable on vis-mig. Ever since the unprecedented influx of this species in the 2017–2018 winter, there seems to have been an uptick of sorts of records of this species, certainly in parts of England. I often find big Redwing days can produce flyover Hawfinches – the finches are known to follow large gatherings of thrushes to pick out seeds from the masses of excrement such flocks produce.

Rarer goodies are possible on vis-mig in October, from nationally scarce species like Richard's Pipit and Lapland Bunting to locally rare birds like Rock Pipit and Woodlark. Two of my best-ever patch finds came on vis-mig watches – Red-throated Pipit and Little Bunting – and it pays to keep an open mind (and have your sound recorder running and camera to hand!). Indeed, Little Bunting is a species to have on your radar during October. Scarce but increasing as a visitor to Britain, it is a possibility both inland and at coastal sites. Make sure you are familiar with Reed Bunting before trying to pick one out of a bunting flock, though it may be more useful to know the 'ticking' call – often the first sign of detection for this subtly beautiful bird and particularly handy during a vis-mig session. Richard's Pipit is another gettable patch find, especially on vis-mig, when the distinctive flight call and large size should be obvious.

It's not just passerines to consider during vis-mig, either. October is an

If you're lucky, a real local rarity, such a Little Bunting, may be picked out in October.

Wildfowl, including Brent Geese, return to wintering grounds during October.

excellent month for migrant Long-eared and Short-eared Owls, as well as other birds of prey like Hen Harrier and Merlin, while wildfowl – including winter swans and geese – are also arriving. On a busy day, a patch vis-mig can genuinely be one of the most thrilling, uplifting and pure forms of birding, as you let the birds come to you and enjoy long-distance migration on your doorstep. It doesn't get much more special than that!

From waders to wildfowl

In a lot of previous monthly sections waders have been championed as a key group to look for. And while that is still the case in October, this is also the time when waterbird attention may shift more to wildfowl. Many duck species migrate south in October and, if you've got waterbodies in your local area, it won't be long before the familiar winter gatherings of Wigeon, Teal and Pochard will form once more. Returning Whooper Swans and Pink-footed Geese will arrive back from Icelandic breeding grounds to their winter ranges – and it's always possible these species may overshoot and arrive in places they don't usually occur.

The same can be said for other species, or you may simply be able to intercept birds heading south and cutting through your area. Brent Goose is a good example – migrant flocks can be picked up almost anywhere, along with rare inland species like Red-breasted Merganser. Turnover at your local waterbody can be daily so it's worth checking regularly if you can. It may well simply be that species relatively common elsewhere might mark a fine local record for you, depending on the context. For me, a Wigeon at pretty much any of my patch waterbodies would be notable – and a Pintail a local rarity!

Down with the dank

There are other reasons to visit wet areas during October. A range of other species are particularly in play at this time of year and sometimes donning wellies and getting out in damp, dank habitats can pay dividends. Snipe return to wintering grounds in October and this is when the first Jack Snipe often come back to favoured sites. With water levels usually lower this early in the season, there tends to be more habitat and, with migrant birds in the mix, it's a good time to encounter this secretive species. If you have a thermal imager then I highly recommend you put it to use – this is a hard bird to see otherwise without flushing it, which you should avoid doing.

Reedy areas are worth checking for Bearded Tit, an irruptive species that may or may not be rare in your area. Both Rock and Water Pipits are on the move in October, with the latter strongly associated with wet marshy habitat. Water meadows can make happy hunting grounds for birds of prey so, with species like Short-eared Owl and Hen Harrier on the move at this time of year, time in these habitats can be productive. Another bird often found close to waterbodies, certainly at inland sites, is Yellow-browed Warbler – though it must be stressed that this 'Sibe' can turn up truly anywhere. It is most associated with the Northern Isles, but migrants on mainland Britain are possible in virtually any location, be it your back garden, local woodland or park. For many patch-watchers, Yellow-browed Warbler is something we all hope for every autumn. In some years there are big influxes, providing a much greater chance of stumbling across one. The first tend to arrive in the Northern Isles or along the east coast from mid-September before filtering through the rest of the country; for many of us, October is the prime time to try and get lucky. They often join mixed tit flock and can be vocal – knowing the distinctive

Yellow-browed Warbler is possible for virtually any patch birder in October, especially in good autumns for the species.

(and surprisingly loud) call is likely to be your best weapon in finding one. As mentioned, they can turn up anywhere, but they fit under this banner as they are often near watercourses of some form – perhaps along a stream or ditch, or in a mixed flock favouring some woodland by a lake, and so on.

Focus on food sources

In September, the weather can often still be pleasant and plenty of insects may be on the wing, certainly early in the month. Thus, it is still a time for insectivores – warblers, chats and so on. As October goes on, the focus on food sources should move to later-season items, such as berries and seeds. At your local woodland site, attention will turn from those open areas that may yield flycatchers to the clusters of fruit-bearing trees holding thrushes, and in farmland areas the hedgerows and livestock fields holding Whinchats and Yellow Wagtails may be less productive, but instead you can check field margins and winter crops for finches and buntings.

Identifying these different food sources is handy for October, but also the winter ahead. If you can find an area of farmland with plentiful seed attracting flocks of species like Chaffinch, Linnet and Goldfinch, then there's a fair chance this will entice other, more unusual species. Any arable land attracting Reed Buntings, for example, could produce a Little Bunting at some stage; see also Yellowhammers and Pine Bunting! Of course, these are rarities and extreme examples, but they are nonetheless areas to seek out. Locally rare birds, like Tree Sparrow and Corn Bunting, could be drawn in to favoured feeding areas. Bramblings, too, will mix with Chaffinches at arable sites, especially if it's a poor winter for beechmast.

Hedgerows are good places to focus on in October, and into November. A good, ancient hedge will be a mix of shrubby bushes and mature trees and, as a general rule, the number of plant species found in a 30m length of hedge indicates how many years old it is. Common hedgerow species include hawthorn, blackthorn, field maple, hazel and spindle, and the best hedges will also include plants such as dog rose, bramble and honeysuckle. In the autumn, their nuts, seeds and berries are a good food source. They also act as a great windbreak, providing shelter during bad weather. Any area, inland or by the coast, with good, mature hedges can provide a great spot to watch.

With the harsher months ahead, you can get ahead of the game and establish where seed- and fruit-eaters are likely to gather in the coming weeks – and regular checking, especially during periods of cold weather, may produce some quality.

Noc-mig

Increasing in popularity in Britain in recent years, noc-mig (nocturnal migration) involves leaving an audio device on to capture bird sounds continuously throughout the night. In October, this can mean recording for 12 or more hours, which makes listening to the entire audio impractical. What makes noc-mig feasible is the ability to transfer long digital sound files to a computer, where they can be displayed as a spectrogram on various pieces of software, including Audacity (a popular – and free – example). A spectrogram visualises how sound frequencies change over time, and understanding it

Redpoll is another finch species that tends to be present in greater numbers from October.

is crucial to the process of identifying the calls you've captured. There is also technology that filters out the calls for you, sometimes even using AI to identify the vocalisations – Birdweather's PUC is one example. If you're a keen patch birder (particularly if you have easy access to safe space where you can leave your recorder) or, better still, are big on your garden list, then noc-mig is a lot of fun.

October is arguably the premium month for nocturnal sound-recording in Britain. Not only is it when rarities and vagrants are most likely to occur, it is also when the largest volume of birds pass overhead. October generally brings the peak passage of thrushes, chiefly Redwing; a good night can see thousands logged. Other species, such as Robin, Blackbird and Song Thrush, can all potentially go over in high numbers, with the large proportion of total annual passage often taking place on just a handful of nights on October.

As mentioned elsewhere in this section, wildfowl are moving in October, and nocturnal listening stations may pick up flocks of Wigeon, Teal and other species on the move, as well as Coot and Water Rail which birders may not typically think of as migrants. Rarities are possible too, though expert analysis and good-quality recordings are always needed to confidently confirm such goodies.

Wherever you live, and whatever your setup, undertaking noc-mig will reveal fascinating insight into what flies over your garden or area after dark. A huge night of thrush passage might include a Ring Ouzel or two, for example, or a waterbird far from suitable habitat might surprise you. It's not just about migrants, either, and noc-mig effort may reveal unexpected resident species like Barn or Little Owls. You don't even need to record things, either – simply sitting outside on a quiet, still night and listening for birds can be hugely enjoyable.

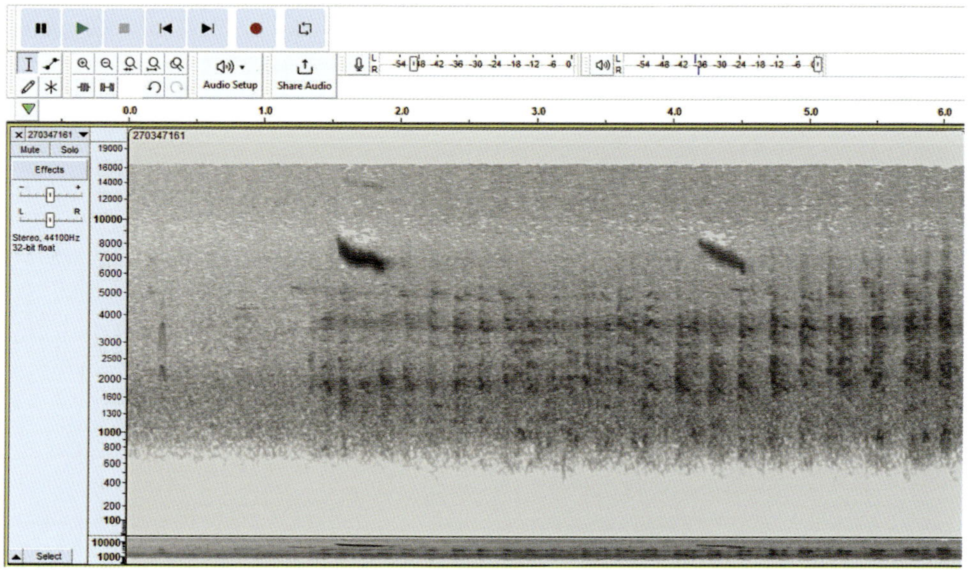

This is a sonogram produced from a recording of a Redwing flock.

11 November

The changing of the clocks in late October means that there is a little period of extra morning light at the start of the November – and thus extra patch opportunity! – but by mid-month the short days are firmly in situ, and winter's grip gradually tightens. Undeniably, at this late stage of the year much in the natural world is slipping towards its winter dormancy. With inclement weather not infrequent, and approaching the end of a long autumn, it can sometimes be hard to motivate yourself to get in the field – but there's plenty of reasons why you should.

November offers a variety of good birding on patch, especially if the weather plays ball. General migration continues well into the month and, by the end of it, true winter visitors are arriving, too. The shorter days make it easier to get out at the first and last knockings of the daylight, too, while birds flocking together ahead of the harsher weeks ahead can produce wonderful spectacles. It is worth one final push before the steadier days of midwinter arrive.

Roost counting

With the sun setting earlier, November offers a great opportunity to watch birds go to roost. I find late autumn dusk sessions to have an almost ethereal feeling to them and, certainly on nice evenings, they're a very relaxing way to end the day. You may well know a spot close to home where impressive corvid or Starling numbers gather – picking a vantage point and watching (and counting) these birds come in to roost can be hugely enjoyable and you might be surprised at how big your totals are.

Corvid and Starling roosts, in particular, are renowned for their dramatic pre-roost displays; many Starling murmurations are well-known, but corvid gatherings less so, even though they can be similarly spectacular. So, perhaps try and find an 'unknown' roost site – follow the flight lines of evening Jackdaws or Rooks and work out where they are spending the night.

Thrushes, pipits and wagtails can also be tallied coming into roost; farmland sites can offer good opportunities for watching birds. Often, such gatherings of passerines can attract predators, with Sparrowhawks and Peregrines among those likely to be lurking. Indeed, raptors themselves can gather for impressive roosts, with Red Kite an obvious example. Other larger species that roost communally include Little Egret and Cormorant.

Wetlands are especially productive places for birdwatching at dusk. You may be lucky enough to have a Marsh Harrier roost in a nearby reedbed, which may also produce a flyby Bittern – a species often active at last light and one that November is a good month for. A chorus of squealing Water Rails, noisy gathering geese and calling passerines make for a great aural experience.

Once the sun has set, some species may become more active and visible. Among them are Woodcock – a bird that arrives in large numbers during November – and an evening beside a damp meadow or arable field might

Right: Watching roosts is a fun way to spend November evenings. If you have a Starling murmuration in your local area it's worth going to experience it.

produce surprising numbers. Lapwing and Golden Plover like to feed after dark, too, and Grey Partridges can often be vocal at this time of day.

Out for owls

In a similar vein, the shorter days mean there's more opportunity to look for owls in November. Areas of rough grassland, river valleys or pasture on your patch are worth visiting late in the day on a bright, clear winter afternoon to look for Barn Owl. Mostly nocturnal, in winter Barn Owls often feed at dusk or even earlier, especially if the weather is cold. It's also a great time of year to seek out Little Owls, too, which can be rather vocal on pleasant evenings. An encounter with either of these two species, no matter how often you see them in your local area, is a treat.

November is also a good time of year for Long-eared and Short-eared Owls. You might be lucky enough to have either of these as a breeding bird, but for many of us they are passage migrants and winter visitors. Late autumn is perhaps the best time of year to stake out Short-eared Owls. This species often has favoured sites but is prone to irruptive years, when birds can turn up almost anywhere. Check any likely-looking owl habitat – areas with long grass and cover for rodents, such as farmland, water meadows, moors and marshes, are the best areas to check. Short-eared Owls can often appear well before sunset too, quite unlike Long-eared Owl, which is far more nocturnal.

Indeed, finding a Long-eared Owl is a much harder task. In suitable looking areas of habitat, such as places with dense, scrubby hedgerows, impenetrable thickets or old hedgerows, be alert to the alarm calling of common passerines – they might just be mobbing a Long-eared Owl. These secretive owls leave white droppings and also regurgitate pellets while at roost sites, so keep an eye out for a scattering of

Shorter days mean it's possible to make the most of dusk, a time when species such as Grey Partridge can become more vocal.

grey thumb-sized lumps on the ground too. If you have a thermal-imaging device then use it methodically in promising places to see if you can pick out a heat signal. Sometimes Long-eared Owls roost in more open places than you might first think – and if one is around, there's a reasonable chance several birds are, such is the communal nature of their winter roosts. It's important to remember that owls are vulnerable to disturbance, so be sure to keep a distance and think before sharing any sightings.

With the shorter daylight hours, late autumn and early winter is an optimum time to search for owls at the end of the day, including Short-eared Owl.

Woodpigeon migration is an underappreciated late autumn spectacle.

Late season vis-mig

Without wanting to repeat advice from the September and October sections, the season for vis-mig continues well into November and can be very fruitful at this time. Milder autumns on the continent are leading to later arrivals of thrushes and finches, so early November can often produce exciting days of movement – and mean patch goodies such as Ring Ouzel and Hawfinch are in play during this month.

November is also a good time for Woodlarks to move around – a fine prize if you don't have them locally – and wintering buntings, like Snow or Lapland. As in September and October, keep an eye on the weather and local bird news channels to get an understanding as to whether or not migration is happening. I have had some days of massive thrush passage in November, sometimes well into the month.

However, in November, it is often Woodpigeons that steal the vis-mig show. Woodpigeon flocks are not typically thought of as a spectacle, but fine, clear (usually cold) days in November with westerly or north-westerly winds can result in thousands migrating overhead. This is arguably one of the most underrated British wildlife spectacles, and it's well worth taking a moment to watch the skies as large flocks pass through.

Early winter storms

November can produce some rotten weather, with wintry blasts from the north often bringing strong winds, cold temperatures and rain. If you live by the coast, such rough conditions can be a good time for seawatching, with storm-driven possibilities including skuas, Leach's Storm Petrel and Grey Phalarope.

November is also the best month for Little Auk. The most frequent time this little alcid is detected in Britain is late October to early November, when birds are travelling south-west from Svalbard to their wintering grounds. High pressure over the UK and a brisk northerly airstream often produces hundreds, occasionally thousands, in northern and eastern coastal areas. Very occasionally huge influxes can occur, one of the most recent being in 2007, which saw several found inland.

Indeed, it is worth checking waterbodies during periods of storms, as covered in the September section. The aforementioned Grey Phalarope has a knack of turning up inland during bad weather, and other coastal species that may not merit much attention on a seawatch but are high value inland – divers, Shag and Gannet, for example – are possible (with a large slice of luck!).

November storms can blow pelagic species like Grey Phalaropes inland.

Working for waterbirds

It may be a cold and wet month of ever-shortening days, but November is perhaps the best time of year for a variety of unusual waterbirds to appear on your patch, with migrating grebes, divers and wildfowl all very much on the cards. On

Below: Following stormy weather, check your local wildfowl, as there may be something unusual like a Long-tailed Duck lurking among the common species.

By November, many geese have returned to wintering grounds, including White-fronted Goose.

those particularly dank and grey days of November, it sometimes feels like it doesn't even get properly light. Yet it is this bleak, murky weather that has the potential to deliver something special.

A range of scarce and localised species are plausible in such conditions, even inland where these birds aren't usually found: think Red-throated Diver, Slavonian Grebe, Scaup, scoters, Long-tailed Duck or Red-breasted Merganser. Late autumn is most likely to throw up a wayward individual, possibly displaced by overnight fog or an easterly blow. Although living near a large waterbody, such a reservoir, will greatly increase your odds of scoring one of these species, they can turn up just about anywhere, including rivers, lakes and even small ponds. But beware that they will often make a brief visit, perhaps lingering for just a single day before moving on.

It is not just stormy weather that should draw you to your local waterbodies during November. Winds from the east, especially coupled with cold temperatures on the continent, can see wildfowl and other waterbirds arrive in Britain in large numbers. Diving ducks are building up at this time and there are plenty of scarce possibilities to look for amid the commoner Tufted Duck and Pochard – think Scaup, Ring-necked and Ferruginous Ducks, or even Lesser Scaup. Sea-ducks can sometimes turn up at this time of year as well, such as Red-breasted Merganser, Long-tailed Duck and Velvet Scoter. Pay attention to dabbling ducks, too, and pan through the common species for possible rarities – Green-winged Teal among Teal, and American Wigeon with Wigeon, for example.

November, especially the end of the month, is also a good time for scarcer grebes. If the wind is in the east, murky or cold mornings are worth checking for Slavonian and Red-necked Grebes – this is perhaps the best time of year to score these patch treats. Divers are firm possibilities during such weather, especially Black-throated and Great Northern.

By the end of November, the bulk of our wintering populations of common geese should have arrived. You may live in an area that has wintering Brent or Pink-footed Geese, for example. If this is the case, then there's a chance of a rarer tag-along – perhaps a Black Brant with Pale or Dark-belled Brent Geese, or maybe Todd's Canada or Richardson's Cackling with flocks of Barnacle, White-fronted or Pink-footed Geese. If you aren't in an area with winter geese, there is a chance of migrating flocks or overshoots being discovered at this time. Fog in the North Sea during late November and early December, coupled with easterlies, will almost certainly send some grey geese over from the Low Countries, and it's at such times that White-fronted Geese might be possible in areas they don't usually turn up, and Tundra Bean may appear in strongholds for wintering geese.

Perusing plovers

By now, both Lapwing and Golden Plover have formed their winter flocks, and many are in areas where they'll remain until the spring. At the end of the harvest, farmland landscapes all across Britain transform – crop fields are ploughed, tilled and reseeded, suddenly presenting these two waders with plentiful foraging grounds. They make for wonderful sights and sounds, and they can also present an opportunity to make an exciting local find, particularly during late autumn.

Depending on where you live, flocks of thousands of Lapwings and Golden Plovers might be a regular seasonal sight in your area. Appreciating them is one thing, but scanning through them for rarities is an altogether more difficult challenge. They are well-known for being skittish and mobile, so panning through a massive flock can feel a bit like searching for a needle in a haystack. Furthermore, various factors – including weather and land usage – means flocks might be displaced from one day to the next.

Perseverance is key – as is taking opportunities when they arise. If you stumble across a flock on one of your farmland sites, make sure to check it as thoroughly as possible there and then – they might be gone tomorrow. And if the flock is flighty, keep working through the individuals when they land again. Of course, you're far more likely to not be rewarded than you are to pick out a rarity – but they are possible. American and Pacific Golden Plovers turn up annually in Britain and many probably go missing in large flocks that birders don't take the time to study. Even greater rarities, such as Sociable Lapwing, are possible. And back towards the commoner end of the dial, Grey Plover has a habit of loitering in large Lapwing and Golden Plover flocks, even away from water, while wintering Dotterels are not unheard of.

Careful scanning through Lapwing and Golden Plover flocks may reveal rarities such as American Golden Plover.

12 December

By December, daylight hours are few, the weather is often miserable and, at the end of a long year, it can be hard to motivate yourself to spend much time on patch. Modern-day winters are often mild and wet, with cold, frosty days at something of a premium. However, there are plenty of reasons to get in the field this month – especially if exciting weather is in the forecast.

That said, after autumn's drawn-out liveliness, December is a good month for taking things at a slower pace and is prime time to enjoy some rest and prepare yourself for the new year – when local birding begins all over again!

Going for gulls

Midwinter is perhaps the best time to spend studying your local gulls. This is when the largest gatherings will occur, be they huge roosts of large gulls on reservoirs, groups of Black-headed Gulls around playing fields, feeding flocks on farmland or loafing assemblages at coastal sites. In winter, a range of species form mixed flocks away from breeding colonies.

Work out where the biggest gatherings occur in your area. If you're blessed with decent-sized waterbodies, whether inland or by the coast, then it's likely that a gull roost is present – this can be a really fun way to spend early winter sunsets, as birds drop in during the fading light. Anywhere with good numbers of large gulls can hold surprises, from local delights such as Caspian or Yellow-legged Gulls to bigger prizes like Glaucous and Iceland Gulls. Inland, Great Black-backed and Mediterranean Gulls may represent decent records – and of course a real rarity isn't impossible! Persistence is important with checking gull roosts.

During the daytime, you might seek out gulls feeding on farmland or other open spaces. It's worth searching pasture or pig fields, if your area has such places, and gulls may also use arable fields that have been left or recently ploughed for winter sowing. Refuse tips and industrial parks are other places to check during the day, as well as any areas of flooding. Gulls will loaf during the day, too, sometimes dropping in to favoured places for short periods. Identifying and knowing such locations is helpful – as is gaining an understanding of the best time of day to visit.

It's worth checking your local gull roosts in the winter for something unusual. Here a Mediterranean Gull is among Common Gull.

Winter storms should prompt you to check your local waterbodies for something unexpected, like Great Northern Diver (left) or Shag (right).

Deep water

December is a time for waterbirds – and particularly those that prefer deep water. At this time of year, flocks of diving ducks gather, and consistent checking of these gatherings can produce rarer birds. Even during periods of unexciting weather, it is worth persevering as turnover of such groups is more frequent than you might think.

The typical carrier species are *Aythya*, chiefly Tufted Duck and Pochard. Any site that holds good numbers of these birds is absolutely worth checking during December. Sea-ducks are possible at this time of year, too, especially in cold weather, with Long-tailed Duck and Red-breasted Merganser on the cards. Smew could also be possible and, for some, commoner species like Goldeneye or Goosander may represent a good find.

Deep-water sites are also the likeliest to attract other desirable inland waterbirds, such as grebes and divers. Of course, these cannot be counted on – but they are absolutely possible at this time of year, especially in the right weather. Of the divers, Great Northern and Red-throated are the most likely inland, with Black-throated rarer. When it comes to grebes, Slavonian is the most typical of the three scarcer species to consider – Red-necked is possible as well, but Black-necked is a less-usual roamer at this time of year.

Cold weather

Cold weather can drive exciting movements and displacements of birds during the winter, especially in December when they will be more inclined to move about than later in the season. A one-off cold day may not do much, but consecutive freezing days can result in bitter conditions on the continent and frozen waterbodies, sending birds roaming around for open feeding and resting areas.

Waterbodies should be the first focus. Some will freeze over quickly, making those that remain open hotspots for wildfowl. At such times, even small stretches of water can have huge groups of birds – and, of course, it's wise to check for rarities, which may have been moved around by the weather. If the cold weather is coming from the east, there is a greater chance of birds moving a from the continent. At such times, grey geese, wild swans and scarce ducks, like sea-ducks and Smew, are possible, along with grebes and divers. Other coastal species, such as Shag, may pitch up inland at such times as well.

Sharp north-easterly blasts can produce Little Auks along the east coast.

Cold weather may drive passerines to migrate to milder areas where food is more readily available, too. As a result, December vis-mig can be a good shout. Thrushes and Skylarks are known for their cold-weather movements, some of which can involve significant counts. A freeze may move around rarer birds, too, like scarce buntings and finches. Lapwing and Golden Plover are also species strongly associated with cold weather, and it doesn't take much of a cold spell for them to start moving around.

Rarely, truly severe weather may occur. A recent example – albeit in February and March – was the 'Beast from the East' in 2018. At such times, it can become very difficult for birds to survive and displacement encounters can be rather grim. Birds such as both snipe species and Woodcock may feed out in the open, with coastal waders forced inland. When such conditions occur, a marked shuffling of the avian pack is essentially guaranteed.

Winter seawatching

It wouldn't be December without a storm or two, and such conditions can be productive for seawatching, should you be lucky enough to live on the coast. Even in steadier conditions, December is a good time to scan the seas for roaming coastal birds.

Alcids, for example, move around at this time of year and persistent scanning may produce a rarer species like Puffin or Little Auk. Scarcer grebes are possible in calm conditions, while stormy days might yield inshore Little Gulls or even Leach's Storm Petrel. This form of birding can always throw up a surprise, so it can be worth giving it an end-of-year go for a late and unlikely tick.

Figuring out finches

Finches flock up during the winter, concentrating at favoured feeding areas. Hawfinches, for example, gather in groups outside the breeding season and, at certain sites, will perch conspicuously on the tops of bare broadleaved trees and conifers, sometimes for long periods. They are closely tied to hornbeam, but also favour areas of yew, field maple, beech, hawthorn, blackthorn and wild cherry. If you can identify a site with such food sources, then you're in with a shout of finding Hawfinches, even if they aren't typically associated with your local area.

Each winter sees different finch species present in Britain in varying numbers – sometimes it can be a bumper season for one species and not the other, and then the other way around the following winter.

If you have plenty of Repdolls locally, look through flocks for rare subspecies, like Mealy Redpoll.

Understanding each species' favoured food source is helpful. Bramblings love beechmast, Crossbills seek out Scots pines, and so on.

Mixed flocks will also gather, especially on farmland sites where there is a reliable source of food. It can be most enjoyable sifting through the assorted species. Furthermore, there is a chance of scarcities or even rarities among flocks of commoner birds. A large Redpoll flock may hold an individual of one of the rarer subspecies, like Mealy or even Coues's Arctic Redpoll, while Parrot or Two-barred Crossbills may be found with Crossbills (albeit very rarely!).

Winter seed-seekers

Wherever in Britain you patch, it's likely you'll have some area of farmland, such is its dominance in the country's landscape. While plenty is degraded and low in biodiversity, there are still areas that provide sanctuary to farmland species. In the winter, some such areas are veritable havens for seed-eating birds, including finches, sparrows and buntings. Many of our much-loved granivorous species have declined massively in recent decades, with birds such as Linnet and even House Sparrow on the Red list of the UK's most threatened birds. However, you can still find pockets of arable farmland that are thronged with such species in the winter months, as they flock together and congregate around areas with plentiful food sources.

Winter food is crucial to supporting populations of such birds at this time of the year. Stubble fields are optimal places to seek out, but winterbird-food plots (areas of sown seed-bearing crops that are left unharvested over winter) can be havens for birds as well. Wheat, barley, oats and triticale are particularly productive. Strips of set-aside can also be attractive. In areas of farmland used for shooting, game cover crops offer food for buntings. The crop used can vary quite a lot, with some rather useless for wild birds (maize being a popular example around my area).

In some areas, typical food sources – such as stubble, areas of set-aside, winter birdseed plots or game cover crops – are not enough, with recent research suggesting that supplementary winter feeding is essential to maintain and restore farmland bird populations.

Despite this, finding such havens on your patch isn't as easy as it was in years gone by. That said, there are still pockets of farmland that do hold impressive winter congregations and they are well worth seeking out, not only to provide some enjoyable birding in their own right, but also to work through for rarer species. Depending on where you live, each species will have a different rarity value – Corn Bunting, Twite and Tree Sparrow might represent major finds, with national scarcities or highly localised birds, such as Cirl, Little and Pine Buntings, also possible.

Depending on where you live, finding such areas can be tricky. If you're rural, it can almost seem like there is too much habitat to work through – searching endless fields for that one pot of gold. Conversely, if you live in an area with little arable farmland, then habitat for seed-eaters can feel hard to come by. Most of the farmland in my local area is used for either livestock or shooting, meaning winter crops are limited, and are often merely small, isolated 'islands' within poor-quality 'pheasant country'. Exploring your local area to find the

In the winter, farmland areas can be magnets for seed-eaters, including Tree Sparrow.

best spots is recommended – Google Maps or an OS map can help. Such patches of habitat are often on land managed by wildlife-friendly farmers and it is worth engaging with any farm worker you might bump into. I have often found they are keen to hear the opinion of a birder. This isn't always the case, though!

If you've found a food source, it's likely it'll have been planted close to a hedgerow. This is key, not just because it offers nearby shelter from predators for the birds using it, but because, from

Areas with cover crops and winter stubble can be attractive to Corn Buntings.

Birding farmland can be productive during the colder days of December.

a birding point of view, this is the best way to scan through the flocks. Patience is required – birds can disappear from view for long periods, and may only sit up for a short time. Finding a discreet, sheltered viewing location overlooking a hedgerow and the edge of the crop is recommended. The birds may not roost in the same place they feed. I find buntings to be fairly lazy risers, and mid-morning is perhaps optimum for working through flocks.

Other factors to consider include the rotation of crops. A field or area that might be bustling with birds one winter can be devoid of them the next, and it's an annual job to get your finger on the pulse of your local farmland site. Seed hoppers may be in place around the edge of some fields, too, especially in areas where shooting takes place, and these can often attract plenty of birds. Indeed, a productive hopper can sometimes offer better viewing than a winter crop, as the birds feed on the ground in an open area.

Time to unwind – and prepare

Despite all December has to offer, there's no denying it's a great time to ease off patch duty somewhat, just as midsummer also offers an opportunity to rest. After the excitement of a new year, a hectic spring and the protracted nature of autumn, one can easily feel a bit 'patch-fatigued' by this point in the calendar. As a result, it can be a great idea to enjoy lie-ins and days without birding, sheltering inside during the worst of the weather.

Furthermore, it can offer time to prepare for January and the commencement of a new birding year. Perhaps you want to outline some patch goals for the coming year – trying for a record year list, taking on a green birding challenge, or focusing on rare breeders. Working out future goals – and celebrating the ones you have achieved this year – are worthy ways of spending birding time during December.

Patch tales: Old ways and new grounds

David Campbell is based in East Sussex, but has spent many years birding in London, Surrey and West Sussex, with patch birding a constant throughout.

The local patch has always been an important part of my birding life. Feeling that I am contributing something useful, even in a small way, is a big part of what makes me tick as a birder. I have always enjoyed travelling to see a rare bird or doing the rounds of known sites for exciting new species, but the deepest satisfaction has consistently come from finding lesser-known local spots and revealing their secrets.

I've had local patches I thought I'd visit for life, but circumstances have led me through many patches over the years, each marking a different chapter. Patch birding took hold before I even realised it: early trips to Carshalton Ponds with my mother evolved into Kingfisher quests up the River Wandle. Grey Wagtail, Snipe and Little Grebe were prize finds that still mean a lot to me.

Later, Nonsuch Park became a fixture. With its mix of gardens, scrub, woodlands and ponds, it felt like multiple hotspots in one. It was where I experienced the thrill of a 'patch tick' for the first time – a Lesser Spotted Woodpecker – and watched breeding Hobbies. Though I missed many migrants back then, it laid the foundation.

Soon, I found myself drawn to Beddington Farmlands, a flagship local site. Though it meant stepping away from Nonsuch, the birding was too good to miss. Only keyholders could access its core, so I relied on long-time patch stalwarts to get in. There I met people whose lives revolved around the patch. Even as a schoolboy, I tried to match their dedication.

Each day brought different possibilities – sludge lagoons attracting Green and Pectoral Sandpipers and even Grey Phalarope. Inland rarities like Arctic Tern and Great Skua felt especially magical. Covering more ground yielded

David has fond memories of patching Banstead Woods in Surrey.

birds like Wryneck and Spotted Crake, while winter meant thousands of gulls and exciting raptor watching. However, I soon felt again the urge to explore neglected ground.

That led me to Canons Farm and Banstead Woods. What started as curiosity soon turned into obsession. I'd bus or drive there at every opportunity. Unlike Beddington, the site was dry, so finding wetland birds became an added challenge I relished. I mapped out species like Little Owl, Marsh Tit and Yellowhammer. Finding a Lesser Spotted Woodpecker nest was a real milestone.

I began treating the patch like a bird observatory, keeping meticulous logs and encouraging others to join me. A community formed: more observers, a blog, pub meets, field walks. Our coverage paid off with finds like Quail, Sandwich Tern and Dunlin. But some of the best moments came in those stolen, low-expectation visits.

One November afternoon, I dashed to the patch after school with barely an hour of light left. It was gloomy, and even I questioned my sanity – until a male Hen Harrier cruised into view. It roosted nearby, and the next morning gave local birders a stunning send-off before continuing east. Two years later, on 4 May 2012 – peak spring migration – I drove to the patch on a college lunch break. With limited time, I joined two others by the farmhouse. We spotted Ring Ouzel, Whinchat and Yellow Wagtails when a shout rang out: "WADERS!" I snapped photos as the small flock circled. When they banked, I saw chestnut bellies and white supercilia: Dotterel! Fifteen of them. It was the first Surrey record in 124 years. We put the news out and dozens visited before dusk. I stayed until nightfall, and a Fox flushed them under the stars.

This 'trip' of Dotterel goes down as one of David's most exciting patch discoveries.

Eventually, I left the patch behind, moving to the Sussex coast. Choosing one patch was impossible with so many excellent sites within 10km — Goring Gap for gulls and seawatching, Cissbury Ring for grounded migrants and the Adur Estuary for tidal variety. Flexibility worked, and on these patches I found Golden Oriole, Hoopoe and Iceland Gull.

Before leaving Worthing, one last October walk at Cissbury brought a buzzy pipit call. I managed record shots of Sussex's third Olive-backed Pipit – the first in 20 years.

Now in Hastings, I've adopted a 'superpatch' spanning Hastings and Rother. With promising terrain and few regulars, I'm enjoying every minute. Picking spots based on weather and season, I've stayed true to what patch birding is about: investing time, building community, being flexible – and always believing something magical could happen.

Short-eared Owl

CHAPTER SIX

Useful tools

Every patch birder needs at least a few tools of the trade. Some are obvious – a pair of binoculars is pretty useful, for starters! But how important is a telescope – or a camera? These days, you can even patch bird at night, with noc-mig having taken off in popularity and thermal-imaging cameras becoming affordable and accessible to amateur naturalists. Furthermore, there are many phone apps that can help you in all sorts of ways, from suggesting which species of bird you're hearing to pointing you towards nearby footpaths. In this section we touch on some of the key kit for the patch birder.

Binoculars

It goes without saying that binoculars are vital tools for any form of birding. When purchasing a pair, the main consideration is magnification. When looking at binoculars online or in person, you'll typically encounter two numbers, denoting magnification and objective lens diameter. For instance, in the case of 8x42, 8x signifies the magnification power, while 42 indicates the diameter of the objective (front) lens in millimetres.

For many newcomers, the feeling is to opt for the highest magnification available – we want to get as close to the birds as possible, after all! However, excessively high magnification comes with image instability. With higher magnifications, even minor movements and vibrations are amplified, resulting in lessened image clarity. Additionally, heightened magnification often affects minimum focus distance, and restricts the field of view.

Consequently, most birders favour binoculars with magnifications ranging from 7x to 10x. Historically, 8x served as the standard intermediary between 7x and 10x. Nowadays, some manufacturers offer magnifications like 8.5x and even 9x, presenting a compromise between the potency of 10x and the stability of 8x. A larger objective lens improves the binoculars' light-gathering power, although adds to their weight.

So, consider the type of birding you do on your patch. Are you up at first light regularly? If so, maybe a larger objective lens size is useful. Will you be carrying your binoculars around your neck while walking many miles? In this case, a lighter pair might be wise. It pays to shop around – visit stores where you can try out different magnifications and models for yourself.

Budget is a final – and very important – factor to bear in mind as well. Top-of-the-range binoculars can cost thousands of pounds, but it's increasingly possible to find more competitively priced models that are of a good quality. Indeed, a solid pair of binoculars can be picked up for a few hundred pounds. It's a good idea to explore the second-hand market, too – shops like London Camera Exchange and CleySpy display a range of used optics that can sometimes be picked up for a snip.

A pair of binoculars are an essential tool for birding.

A telescope can be very useful when scanning waterbodies and large vistas.

Telescope

Telescopes are another useful tool, though their suitability for patch birding can vary – if you are walking or cycling, their size and weight can make them awkward to carry around.

The questions to ask are the same as for binoculars – what level of magnification are you aiming for? What's your budget? Additionally, how much weight can you comfortably carry around?

Most telescopes provide magnification ranging from 15x to 60x, with speciality scopes offering even higher magnification. You can opt for a telescope with a zoom lens or interchangeable eyepieces for variable magnification. Fixed-lens telescopes typically offer 20x or 30x magnification, with a 40x lens suitable for long-distance viewing.

Another decision to make is whether to opt for a straight or angled telescope. Straight telescopes have an eyepiece aligned with the objective lens, making them easier to use when in a hide, for example. They are typically more budget-friendly. Angled telescopes feature an eyepiece set at a 45° or 90° angle to the scope's body – this helps to track moving birds and reduces neck strain, particularly for taller people. With an angled scope, you can position the tripod lower, enhancing stability. Ultimately, it comes down to personal preference.

As with binoculars, the cost can vary, but generally telescopes are expensive – and tripods can be too, with premium tripods costing as much as a budget pair of binoculars! As a result, it's important to consider how much use you might get out of a telescope on your local patch. I find that I don't use my 'scope much while patch birding. It's useful for scanning the larger waterbodies and for studying breeding raptors, but for probably 80 per cent of my local birding I don't use or even carry it with me. Transport is a large reason why – it is simply not efficient to lug a heavy 'scope around when on foot or bike. But the importance of a telescope will vary massively depending on where you live. I have birding friends for whom the daily usage of a 'scope is essential, such are the large, expansive sites they bird. So, take time to figure out how useful a telescope will be for you.

Camera

Photography is a massive element of birding these days, however you look at it. Bird photography has increased in popularity markedly in the last decade or so, as accessible, affordable digital kit has allowed more people to take part. A camera is a very useful tool for the patch birder, too. Traditionally, field sketches and detailed notes would form an intrinsic part of identification – but a few shots from a camera can capture features within milliseconds.

My personal take is that a camera is pretty much a 'must-have' for the patch birder, and for various reasons, namely:

- **Documenting rare or unusual birds** It's not always possible to do this, of course, especially if a brief fly-by or distant bird among a throng of others is the subject. But more often than not, if you've had time to safely identify a rare bird, you have time to get a photo (and often even a poor-quality photo can be sufficient to nail an identification).
- **Learning** I have spent many hours poring over photos I've taken of birds I've not been able to identify – trickily plumaged gulls, for example. The educational benefits of photos can't be underestimated, and sitting at your computer with field guides by your side is a good way to learn 'post-field'.
- **It's fun!** Bird photography is both enjoyable and challenging, and when you nail a good photo of a bird – regardless of how rare or common it is – you feel a sense of satisfaction.

I would go as far as to say that, if you're keen to either become a skilled patch birder or improve your current efforts, carrying a camera is, in this age, essential. I know few patch birders who don't carry one. There is an argument of 'time spent taking a photo deducts from time spent enjoying the bird', but

A camera is incredibly valuable when it comes to birding.

in practice that's rarely the case, unless the encounter is brief (in which case you wouldn't have had much time to enjoy it anyway!).

Camera technology is continuously progressing – and the range of camera types is far wider than that of binoculars or telescopes. Bridge cameras are the entry-level type, but most birders favour DSLRs with a telephoto lens. Increasingly, mirrorless cameras are being used, too, and these will likely dominate the top end of the birding market in years to come. Weight and portability are important factors. Bridge cameras are generally compact and easy to carry around, while a DSLR with a (necessarily) powerful birding lens can be heavier and clunkier.

Bridge camera quality can vary and, with that, the cost. There are some superb models available, though, and for competitive prices – as with optics, shopping around and trying out different options is important. If documenting the birds you see on patch is your priority, and you're less fussed about top-notch bird photography, then a bridge camera is probably the route to go down – generally, they are lighter and cheaper than DSLRs. The latter, however, are more powerful overall, with image quality and reach generally higher. If you're in the field a lot, this is probably the way to go.

Finally, as touched on elsewhere, if you have a camera, it's wise to have it handy at all times – anything could appear or fly over at any point, and might not hang around for you to reach into your bag! Having a good-quality strap means that you can carry your camera hands-free while you bird. I use a shoulder strap – ideal for both weight distribution and quick access.

Sound-recorder

During the last decade or so, the value of sound-recorders as essential field tools has become more recognised than ever before in birding. The explanation has, largely, been twofold: nocturnal sound-recording (for noc-mig), which has steadily become more 'mainstream', and people carrying sound-recording gear in the field, whether to specifically record birdsong and calls or to try and document a rare or unidentified bird. There are plenty of rare species – including potential patch goodies – that require recordings to support or even fully confirm the identification (Iberian Chiffchaff and Eastern Yellow Wagtail are two examples).

Fundamentally, noc-mig recording involves leaving an audio device running continuously to record a full night's worth of sounds (see pages 143–145

Sound recorders are useful for identifying birds by their songs and calls, documenting rare species and capturing nocturnal migrants (noc-mig).

for more detail on noc-mig). Noc-mig is fascinating and reveals species that would otherwise go undetected. It is time-consuming, though, and of course you can't tick what you don't hear/see! Daytime recording is valuable too. One option is to continual sound-record (CSR). This is rather like 'noc-migging' during the day; you keep a recorder running while you go about your birding, in order to document every song or call heard while you're in the field. This requires fixing the recorder to your person, a rucksack or hat, which isn't always that easy... but it means you won't miss that flyover Tree Pipit or Hawfinch! Alternatively, you can carry your recorder in your bag or pocket then turn it on as and when you want to record something. Some apps, like the RØDE Reporter app, can make for useful alternatives to a physical recorder as well.

Sound-recording is like photography – you can get surprisingly good results of close subjects with a smartphone, or you can spend a fortune on dedicated kit that will extend the range and increase the quality of your recordings. Without an attached microphone with a parabolic reflector, sound-recorders are small and portable, too, making them easy to carry around. Olympus, Tascam and Zoom are popular brands with birders. Audiomoths and PUCs are increasingly used, too.

Thermal-imaging camera

Thermal imaging is one of the newest additions to a birder's arsenal of equipment, using heat signatures to detect birds that are invisible to the naked eye. The technology uses medium- and longwave infrared radiation to create a heat image (known as a thermogram) of the view in front of you. The image displays the temperatures as either different colours or shades, or as a monochromatic image. It captures temperature variation relative to the objects within its field of view to create a picture – hence this may alter as you pan and bring hotter or colder areas into view. For instance, a colder sky or hot rocks will cause the entire image to change depending on the relative heat of what else is present in the field of view.

Costs can be high for a thermal imaging camera and the more you pay the higher the quality will be, both in terms of image resolution and detection range. That said, the size of units has

A thermal-imaging camera is excellent for locating nocturnal species, such as owls, as well as more cryptic birds. Here, a Eurasian Nuthatch can be seen visiting a nest hole.

Thermal-imaging cameras have revolutionised the ability to locate cryptic species like Jack Snipe.

decreased and the affordability has increased to a point where there are now pocket-sized scopes that start at less than £1,000. For many it might be an unnecessary luxury or a piece of kit too far when added to bins, scope and camera. However, there's no doubt that certain 'classic' patch species are much easier to find with a thermal imager. Jack Snipe is perhaps the most obvious example, along with Woodcock and owls. Nocpix, Pulsar and Zeiss are popular brands.

Apps

These days, there is an app – or multiple apps! – for everything. This includes for recording bird sightings, sound-recording, tracking weather, finding walking routes and so on, and equipping your phone with various ones is very

The app essentials

Below is a list of just some of the apps that offer the most use to the patch birder. Some are free while others have a small cost.

BirdTrack The BTO's free and easy way of logging your bird records online.

Ebird Similar to BirdTrack, a free resource to create and submit lists of birds you see.

Collins Bird Guide An app version of the popular field guide with some additional features, such as being able to compare different species, bird songs and calls. Many other field guides are also available as apps.

Merlin Bird ID Powered by eBird, this AI-supported app helps you identify birds you see and hear.

BirdGuides The leading bird news service in Britain, providing subscribers with real-time updates and exact locations.

OS Explorer The official OS Maps app, providing up-to-date routes for running, hiking, cycling and walking.

Google Maps Provides real-time GPS navigation, traffic and transit information, as well as satellite imagery (ideal for prospecting new patches!).

Strava Tracks your walks and cycle rides via GPS.

Met Office Weather Accurate, detailed weather forecasts throughout the UK, including visibility and precipitation percentages, from one hour to seven days ahead.

BBC Weather The latest weather forecast from BBC Weather. Easy to use, with hourly forecasts.

Windy Provides a live weather radar, detailing wind, rain and much more, with an animated weather map.

useful. As mentioned elsewhere in the book, there is no need to carry a notebook, although of course you may wish to anyway. Now, you can enter your notes directly into an app, and they will then go into a database of all your bird records.

Books

Despite many field guides now being available as apps, which is very useful when in the field, it still pays to have a decent selection of titles at home for reference, especially when going back over photos and notes. Some recommended titles are listed in the box on the right.

Bag

It might sound fairly obvious, but a good-qualityrucksack is a valuable patch birding tool. You'll experience all sorts of weather on patch, so keeping your optics and camera dry is important – and having extra layers in the winter or a rain jacket during showery conditions is always handy. And, of course, it means room for carrying snacks! Choose something reasonably weatherproofed, with accessible compartments, and that you can comfortably wear for long periods.

Bike

It would be remiss to not touch on the value of a bicycle for patch birding. Depending on the size of your patch, you may do most of your birding on foot from home. Alternatively, you might drive a car or take public transport. Having a bike, however, is a great – and green – option. I often use a bike, as it's ideal to take me the relatively short distances between the various sites in my patch. It's also great for checking the 1km patch.

I ride a hybrid, as I don't tend to carry a telescope and, often, I'm negotiating footpaths and bridleways as well as roads. However, it is quite heavy and not very comfortable for longer distances on the road, nor well-suited for carrying birding kit. So, depending on where you live, what you're carrying and how far you'll be cycling, a road bike might be a better fit. You may find yourself experimenting with various ways of carrying your gear before settling on the right balance.

Other bits

There are various other accessories that can be useful to the keen patch-watcher. You may want a thermos flask and a sturdy lunch box, as well as appropriate footwear depending on the terrain you're covering, clicker counters and a foldable chair if you do lots of seawatching, and so on.

Equally – and such is the beauty of local birding – you technically need none of the tools cited in this section. All you need to do is get outdoors, with your eyes and ears, and enjoy what birds are present in your area.

The birder's bookcase

Britain's Birds (Hume, Still, Swash, Harrop and Tipling; WILDGuides; 2016).

Collins Bird Guide, third edition (Svensson, Mullarney and Zetterström; Collins; 2023).

Europe's Birds (Hume, Still, Swash and Harrop; WILDGuides; 2021).

Flight Identification of European Passerines and Select Landbirds (Cofta; WILDGuides; 2021).

Flight Identification of Raptors of Europe, North Africa and the Middle East (Forsman; Helm; 2016).

Gulls of Europe, North Africa, and the Middle East (Adriaens, Muusse, Dubois and Jiguet; Princeton University Press; 2021).

Helm Guide to Bird Identification (Vinicombe, Harris and Tucker; Helm; 2014).

Identifying Migratory Birds by Sound in Britain and Europe (Wroza; Helm; 2024).

ID Handbook of European Birds (van Duivendijk; Princeton University Press; 2024).

Waders of Europe, Asia and North America (Message and Taylor; Helm; 2020).

Patch tales: Patience and happenstance

Amy Robjohns is based in Hampshire. She is a strong advocate of low-carbon birding and is a devout patch-watcher.

Amy fondly recalls finding a Whiskered Tern in May 2020.

My local patch includes Titchfield Haven National Nature Reserve on the Hampshire coast, a mosaic of reedbeds, scrapes, water meadows and the mouth of the River Meon. It stretches from farmland south of Titchfield village to the seafront at Hill Head, and follows the canal path up the valley. It's not the best seawatching site, but spring passage still brings skuas and terns, and the patch has offered a wealth of memorable moments – daily changes, seasonal shifts and rare finds that stay with you.

One of those moments came on 27 May 2020. I overslept and rushed to the canal path, hoping the tail end of spring might still hold a surprise. The usual warblers were singing when suddenly I heard an elaborate stream of mimicry – snippets of Goldfinch, Swallow, Nightingale and more, all from a Marsh Warbler. It was only the second I'd found on patch; the first, in 2017, had been unmistakable. Reed Warblers may mimic, but not like this.

I didn't have time to reach the seafront, so instead I checked the footpath near Posbrook Flood. A tern flew past, oddly dark below, pale above, with a short, forked tail. It circled the flood like a Black Tern – but it was a Whiskered Tern! My first ever, and on patch – before 7:30 am, no less.

Rarity-finding often comes down to being in the right place at the right time—and knowing where to check depending on weather and season. Sometimes, circumstances help too. A thick fog in May 2016 ruled out seawatching, so I turned inland and found a Caspian Stonechat near the reserve entrance. A spring Red-rumped Swallow in 2020 came courtesy of lockdown and a later-than-usual walk. The first patch Glossy Ibis in five years flew over in 2019; now they're regular. Patch birding reveals these subtle changes over time – some hopeful, others less so, as climate and habitat shifts bring both winners and losses.

Migration days are rewarding even without rarities. In April 2018, a misty day brought Wheatears, Swifts, a Grasshopper Warbler and an evening Wryneck. The next morning began with a singing Wood Warbler. That year was my best to date, ending on 180 species including Little Stint, Long-tailed Duck, Red-necked Grebe, Manx Shearwater and Snow Bunting.

Over the years, I've expanded the patch slightly to include farmland near Posbrook. It added variety and a new set of surprises. One spring day brought 14 Wheatears in a single field, along with Whinchat, Black Redstart and Lesser Whitethroat. In autumn, the fields attracted Yellow Wagtails, Meadow Pipits, warblers, and flyovers like Spotted Redshank and a Wryneck. That same evening I found a White-winged Black Tern offshore.

Of course, not every session brings a highlight. There have been plenty of dips – like the Buff-breasted Sandpiper in 2016 that flew out of the reserve just before I arrived, or the mystery warbler in 2017 that looked like a Two-barred, but vanished before I could get a photo. Those moments remind us that birding is about patience as much as excitement, and that rarities really are rare.

This Caspian Stonechat sparked a major twitch on Amy's patch.

Ultimately, patch birding is about perspective. The highs are heightened by the quiet days. Watching a place through the seasons teaches you how birds respond to weather, to habitat, to time. You learn where to look, what to expect – and when something unexpected changes everything.

Falls of migrants, like Wood Warbler, in spring and autumn stand out for Amy as exciting patch moments.

Whimbrel

CHAPTER SEVEN

Patch wellness

Time spent in nature can greatly enhance our mental wellbeing. The connections between nature and mental health are well documented and, generally, spending time outdoors is considered to be highly beneficial. However, every so often birding can generate negative feelings – especially when you're a dedicated patch birder for whom the hobby is a part of day-to-day life. As with many things, balance is key, and it's important to keep a check on your mental health even when pursuing a hobby that, nine times out of 10, brings joy and happiness.

Finding balance

Over the years, many revered naturalists have battled with mental illness and found comfort in their passion for wildlife. Several have written about this. For many people, including myself, birding is about forging a deeper connection with birds – observing and understanding their behaviours and appreciating their links with seasons and places. It is also, without a doubt, a form of escapism – a word that often comes up in conversations with birding friends and colleagues. When we're in the field, we can feel at peace and almost as if we're inhabiting a different world. A visit to a patch typically has a calming effect (sometimes an exciting one!) and I've no doubt patch watching helps keep other elements of busy lives balanced.

Birding, and patch watching, also help provide focus and drive. I'm sure I'd have had many more spring lie-ins without it! It's something that can only be undertaken with suitable levels of passion and enthusiasm – but it's not hard for that to teeter over the brink and into obsession and fixation. Add in the pressure some may feel to achieve big lists, see all the 'trending' rarities and find good birds, and then the hobby may start to feel something of a chore.

Time spent birding in nature, even observing more commonly seen birds, such as Robin, is worthwhile.

When birding becomes a burden

I've met – and know of – some people for whom obsessive patch or local birding became too much. One famous former Surrey birder quit his job to chase county and local year lists, but when someone else saw a Long-tailed Skua fly over his patch, while he was on site, he couldn't take it. That evening, he deleted all of his social media accounts, resigned from the county bird club records committee and was essentially never heard of again. I also know of birders whose sole motivation is to find rarities. By struggling to do so over long periods of time, they drive themselves into deep pits of frustration. By its nature, birding can attract more single-minded and eccentric types of characters, and sometimes the perfect storm can brew... I'm sure we've all heard tales of patch-watchers driven to the point of exasperation by their dedication.

It's also worth remembering that, on the odd occasion, patch and local birding can feel monotonous. You will undoubtedly have days where you see nothing of note – at times, this can go on for days or even weeks. Furthermore, frustration with your local patch may materialise in the form of other people, usually non-birders. These people will likely be using the patch for other reasons – where I live, dog walking is particularly popular, and most years I have several rather miserable encounters with a dog off a lead chasing birds or groups of them being loud and disruptive. Such

Visiting your patch can provide a welcome break from hectic modern-day life, though equally it can create frustrations, for example if you've witnessed some disturbance to birds or habitat destruction.

moments can leave you feeling annoyed, but it's worth trying to engage with dog owners in a friendly way – or with other people inadvertently disturbing birds – as I have often found people to be understanding and cooperative (if approached nicely!).

Ultimately, though, patch birding brings a sense of connection and harmony that can greatly enrich your day-to-day life. Watching birds play out their lives in your local area is a beautiful thing and, even if you are having a flat time, the next exciting or uplifting moment is just around the corner!

Trivial issues

It would be easy to say that frustrations and feeling of disillusionment due to birding is a 'First World problem', especially if you're actively choosing to invest so much mental (and physical) energy in the hobby. And, of course, I'm sure all of us view our birding with different degrees of seriousness.

The reality, however, is that emotions can seep into the hobby – especially 'in the moment', and if you are someone who struggles with your mental health already, it is surprisingly easy to find yourself in a rut.

You can end up in a situation where a negative patch mindset becomes self-fulfilling, and you feel like you're forcing yourself out but with the feeling that you're not going to see anything, because your luck is out. That is not what birding is about – and when it feels like it's becoming a chore, it's time to take a break.

Birding – or patch – fatigue is something to be alert to. When you feel like you're flogging the proverbial dead horse, perhaps it's worth taking a step back. Birders who know their patches well will recognise when things are 'out of form', be it a run of poor weather conditions, bad habitat state or simply a lack of birds. These are the moments to learn to identify – and learn to use to take a step back.

Momentum

Momentum plays a big role in patch birding. The aforementioned quiet spells can easily lead to a downward spiral of frustration and a defeatist attitude. Similarly, if you've missed a couple of good birds in your area, it might seem like you've lost your magic touch.

However, momentum so easily swings the other way. One good bird, or one encouraging sign of migration or breeding, and things can feel better. Ultimately, it's those moments that make birding such a fulfilling hobby – and even if they are few in number, they can alter the perception of a day or season on patch. Such are the fine margins of looking for birds, which is akin to panning for gold – sometimes just a few small nuggets are what you need to keep going with a positive mindset. Often, I find, one good patch moment soon leads to another, and before long you feel like you're building up a real head of steam, and go into the field full of enthusiasm and acutely tuned in.

I remember feeling this way ahead of one of my best patch finds, a Short-toed Lark in September 2020. For the preceding month, I'd enjoyed a growing run of decent discoveries, particularly in the passerine department. It almost felt

Patch birding is about creating a deeper connection with wildlife.

like they grew in quality, too, starting with a mid-August Pied Flycatcher, before really picking up with an early September Wryneck. By mid-month I was full of positivity and energy – and that's when I found the lark.

A positive attitude undoubtedly helps – and it's much easier to feel upbeat when you're on a good run. When you are feeling 'in the zone', you find yourself willing or even expecting something good to appear. This heightens your senses, sharpens your concentration and, often, results in success, as it makes you more likely to pick up on even the subtlest of signals. On the contrary, negativity has the opposite impact, dampening the senses and leaving you disengaged and less sharp.

Ultimately, finding and seeing good birds requires a degree of fortune and plenty of effort. Only you can put in the hours or the miles that will shorten your odds of connecting with a rarity. But you can put in all that effort and still come away empty-handed; you need the birds to do their bit and pop up in front of you!

This Short-toed Lark was found during a period of good birding form.

Balance

The key thing with all of this is to maintain a sense of balance. It's perhaps the main reason why I choose to patch a local area, as opposed to a single site – I'm sure I'd find myself frustrated far more regularly if I was birding the same place all the time (although that's absolutely not to say it can't work for others).

When it comes to patch birding, don't underestimate the power of change; just a day's session somewhere different can work wonders and reset your birding mojo. If you're feeling at a low ebb, why not check out a 'new ground' site, or visit one you barely go to. Or simply up sticks and vacate the patch temporarily – it can feel so refreshing to be not just birding a different place, but a totally different region.

Know when to take a step back, too. That application of restraint is key to not sending yourself down a negative spiral. While setting targets and challenging yourself to be better is healthy, the pendulum can swing too far the other way and it is important not to become too embroiled in negativity when the inevitable lean spells come. Remember that birding has a way of levelling itself out over time and any disappointments will give way to periods of joy and fulfilment.

The ups and downs of patch birding and bird finding are what makes it so fun. Its unpredictable nature is why it appeals to so many – the bird of your dreams could appear in front of you at any time and in any place. It's important to manage intensity levels and avoid patch fatigue. Sometimes taking a step back is the answer – it is just a hobby after all! Simply being outside and enjoying common species is a joy itself.

Patch tales: Riding the rollercoaster

Josh Jones is based in south Lincolnshire. He has birded his whole life, across Britain and beyond, but finds patch watching the most rewarding and meaningful way to embrace the hobby.

Some of Josh's earliest patch memories involve seeing flocks of winter Smew.

The foundations of my birding are deeply rooted in watching a patch. I grew up in Langtoft, south Lincolnshire, on the edge of the fens. Here, a variety of habitats – especially the old gravel and sand quarries – form a rich patchwork of waterbodies, perfect for nurturing a young birder.

My earliest birding memories from the 1990s involve trips with my father to these gravel pits. I remember the thrill of seeing Smew glowing white on dark winter waters, and the anticipation of watching new excavations for returning Sand Martins each spring. In May 1997, I witnessed nearly 40 Black Terns in one flock – a moment that left a lasting impression.

By secondary school, birding became a real obsession. I started birding solo at the gravel pits, learning through repetition and slowly sharpening my skills. Patching became my cornerstone. It gave me a deep understanding of birds that twitching alone could never provide.

The highlight of my teenage years came on 25 September 2007. I was birding locally when I heard a sharp, unfamiliar call. A wader flew overhead, then circled and landed. It gave a Greenshank-like *chew-chew-chew* but lacked the distinctive white wedge of a Greenshank. Its yellow-orange legs and slightly upturned bill revealed I was looking at a Greater Yellowlegs – the rarest bird I've ever found in Britain. It stayed just 20 minutes before vanishing, reappearing the next day on the Hampshire coast.

After university in Sheffield and nine years living in London, I missed patch birding deeply. The Thames offered gulls, but it wasn't the same. When the COVID-19 pandemic hit, I packed up and returned to Lincolnshire. Unexpectedly, this move rekindled my relationship with the gravel pits. Daily walks during lockdown reminded me just how special patch birding could be.

In the years since, I've spent countless hours walking and cycling my local area. There is no substitute for time in the field. With tools like eBird, my birding has become more structured and data-driven. Recording full checklists allows me to track trends, monitor migration timing and anticipate what species might appear based on past patterns. I've also embraced weather-watching – being tuned in to the forecast has paid off many times.

Keeping an eye on national news through platforms like BirdGuides helps too. If I see a spring influx of Black Terns elsewhere, I know to check my local lakes. Similarly, the autumn 2024 Hawfinch reports inspired me to try some vis-migging – and it worked.

That said, not every season is productive. In late 2022, after a slow autumn, I took a weekend trip to Belgium. While I was away, a Red-flanked Bluetail was found just a few miles from home. I missed it – the bird vanished before I returned. The disappointment lingered for months, made worse by a sluggish spring in 2023. A brief glimpse of a Roseate Tern on 9 May finally raised my spirits.

Josh has found hours of enjoyment birding in the Lincolnshire fens.

Then came a dramatic turnaround. From late May to mid-June 2023, I recorded a run of good birds: Black-winged Stilt, Little Stint, Caspian Tern, Honey Buzzard and a Purple Heron, among others. After a barren spell, the patch was suddenly alive with possibilities. That electrifying stretch reaffirmed a key truth: birding evens out. Quiet periods are tough, but perseverance brings eventual reward.

One of my proudest discoveries came after such perseverance. In autumn 2021, I spent weeks scanning Golden Plover flocks in the fens. Conditions were frustrating – the birds stayed distant, the weather was rough, and I was getting nowhere. But on 6 November, I stumbled upon 2,000 plovers near the road. Amid them stood one cream-coloured bird: a Pacific Golden Plover. After hours of patience, it finally showed well, even flying right over me, offering clear photos and a diagnostic call. It was a classic case of preparation, grit and just a little luck.

Of course, patch birding isn't all about rarities. The joy also lies in building an intimate knowledge of your local area and its regular species. I've found breeding Long-eared Owls, heard Nightingales singing in spring and confirmed the presence of breeding Turtle Doves – magical, personal moments that build your bond with a place.

Patch birding teaches patience. It can be weeks between standout birds, but each dry spell makes the highlights shine brighter. The real satisfaction isn't just in the birds you see – it's in the stories you gather, the intuition you develop and the thrill of finding something special on ground you know better than anywhere else.

Garganey

Glossary

Birding terms

Alcid Auk.

***Aythya* – also see Diving duck** Genus of diving ducks that includes Tufted Duck and Pochard.

BBRC – British Birds Rarities Committee, which assesses submitted records of rare species for acceptance or rejection.

Barwit Bar-tailed Godwit.

Blocker A bird that's particularly hard to get on a certain list. For example, Pallas's Sandgrouse on many keen birders' British lists is a blocker; they can unblock Pallas's Sandgrouse by seeing one.

Bush-bashing Systematically checking through scrubby habitat for passerines.

Carrier species A gregarious and relatively common species whose flocks may be joined by rare species. For example, Wigeon is the usual carrier species for vagrant American Wigeon.

Common passerine – also see Passerine Unless otherwise stated, and depending on the context, but generally the likes of Blackbird (and other common thrushes) Wren, Dunnock, Starling, House Sparrow, common finches (Chaffinch, Goldfinch, Greenfinch), common tit species and Pied Wagtail.

Common woodland species Generally the likes of Blue Tit, Great Tit, Coal Tit, Long-tailed Tit, Goldcrest, Nuthatch, Treecreeper and Great Spotted Woodpecker.

Corvid Birds of the crow family.

Dabbling duck Duck species that feed by dabbling just below the water's surface, or upending, and which generally favour freshwater, such as Mallard, Wigeon, Teal and Shoveler.

Darvic ring A type of coloured leg ring used for bird study. Darvic rings are large with a contrasting letter and number code that can be read in the field.

Dip Not seeing a bird you've attempted to twitch (see **twitch**).

Diving duck Duck species that dive below the water's surface to find food, such as Tufted Duck, Pochard, Scaup and Common Eider.

Fall A mass arrival of migrant landbirds grounded by adverse weather conditions.

Female-type A bird in plumage typical of an adult female of its species (which in some cases also includes sub-adult males).

Gamebird Birds of the order Galliformes, many of which are 'quarry species', including partridges, grouse, Pheasant and Quail.

Genus A group of closely related species. Closely related genera grouped together form a family.

Gos Goshawk.

Greatspot Great Spotted Woodpecker.

Grey geese Geese of the genus *Anser*, generally referring to any species other than Greylags.

Gripped To be jealous or in awe of a bird/birds another birder has seen.

High summer Colloquially speaking, the hottest part of the summer, but in birding parlance it is generally used to refer to the slight lull between the end of spring migration and the start of return passage.

Hirundine A member of the swallow or martin family (Hirundinidae) – in the British context, Sand Martin, Swallow and House Martin.

Insectivore In the context of this book, any bird species that has a diet consisting largely or entirely of insects, such as warblers and hirundines.

Landbird Bird species that spend all or most of its life on or over land (though may migrate over the sea).

Large gull Larger species of gull, including Herring, Lesser Black-backed and Great Black-backed.

Lesserspot Lesser Spotted Woodpecker.

Mega An extreme rarity.

Migrant Any bird that regularly migrates to and from – or through – the region.

Migrant chat Any routinely migratory member of the chat subfamily, including Wheatear, Whinchat, Nightingale, Redstart and Black Redstart.
Mipit Meadow Pipit.
Noc-mig Nocturnal migration.
'On the deck' When a bird is seen on the ground (or perched or swimming), as opposed to flying over.
Oyc Oystercatcher
Passage – also see Passage migrant The movement of migratory bird species, particularly in spring and autumn.
Passage migrant Birds that routinely migrate through a given area but don't stay to breed. Examples in Britain include Ring Ouzel, Whinchat, Pied Flycatcher and various wader species.
Passerine All songbirds – members of the order Passeriformes, which have feet adapted for perching; includes corvids.
Patch gold A species generally considered common in Britain, but particularly – and quantifiably – rare at a certain patch or site.
Peg Peregrine.
Raptor Generally refers to diurnal birds of prey (so including hawks and falcons but not owls), such as Sparrowhawk, Common Buzzard and Kestrel.
RBBP The Rare Breeding Birds Panel, which defines rare breeding birds and collates breeding records for them.
Rarity Generally denotes a rare bird for which sightings of some species are referred to the British Birds Rarities Committee for vetting (BB rarity), but also for birds at a county or local level.
Redhead A female-type sawbill duck – Goosander, Red-breasted Merganser or Smew, which have reddish-brown heads in this plumage.
Ringtail A female-type Hen, Montagu's, Northern or Pallid Harrier.
Roding Territorial flight of Woodcock, usually at dusk and dawn.
Sawbill Collective term for Goosander, Red-breasted Merganser and Smew (and their overseas relatives); diving ducks that prey on fish and have serrated bill edges.
Scarce grebe The three less common members of the grebe family to occur in the British Isles; Slavonian, Black-necked and Red-necked.
Scarcity Species that, while perhaps not qualifying for vetting by regional and national rarities committees, are nonetheless noteworthy in the region or at a particular site.
Seabird Refers to birds of the open sea that, under normal circumstances, only come close inshore to breed. Includes shearwaters, storm petrels, Fulmar, Gannet, skuas, Kittiwake and auks.
Sea-duck Duck mostly confined to the sea, including Common Eider, Long-tailed Duck, scoters and Red-breasted Merganser.
Seawatching Persistent scanning of the sea from land to observe birds moving past the coast.
Spotfly Spotted Flycatcher.
Sprawk Sparrowhawk.
Thermal Rising body of warm air that often forms between a hillside and lower ground, on warm days. Used by raptors and other soaring birds to gain height with minimal effort.
Tubenose Any seabird of the order Procellariformes, from storm petrels and Fulmar to albatrosses.
Tuftie Tufted Duck.
Twitch Travelling any distance to see a particular individual scarce or rare bird found or reported by somebody else. A birder who is especially committed to this activity might describe themselves as a twitcher.
Vagrant Aspecies outside of its typical range, normally a rarity.
Vis-mig/Visible migration The observation of the diurnal movement of birds.
Warbler A term for various groups of small, insect-eating passerines, most of which are migrants and noted for their distinctive songs.
Warbler, *Acrocephalus* A genus of warblers normally associated with reedbeds and marsh vegetation, including Sedge and Reed Warblers.

Warbler, *Phylloscopus* A genus of warblers, also known as 'leaf warblers', which are normally associated with woodlands, including Chiffchaff and Willow Warbler.
Waterbirds – also see Waterfowl and Wildfowl All birds associated with wetland habitats.
Whitefronts White-fronted Geese.
Waterfowl – also see Waterbirds and Wildfowl Refers to wildfowl and some other birds associated with water, such as divers, grebes, Cormorant, herons, rails.
White-winger Colloquial term for the scarce white-winged gulls; Glaucous and Iceland Gulls (but generally not Mediterranean Gull).
Wildfowl – also see Waterbirds and Waterfowl Birds belonging to the family Anatidae; swans, geese and ducks.
Winter swans Bewick's and Whooper Swans.
Wreck An arrival of storm-blown seabirds on the coast or miles inland.

Habitat terms

Brackish Refers to the presence of salt water in estuaries, where seawater mixes with freshwater.
Chalk grassland/downland A rare and declining habitat, only found in north-west Europe, where shallow soils on chalk bedrock are kept open through grazing and human management. These habitats support some very specialist plant and invertebrate species.
Floodplain – see also Marsh Low-lying land in a river valley, prone to flooding – and often managed intentionally for this purpose.
Headland An area of coast that projects further into the sea than the surrounding coastline, making it the first landfall for birds crossing the sea in that area.
Marsh – see also Floodplain: An area of low-lying land that floods in wet weather or at high tide, and remains waterlogged for long periods of time.
Mire Any area of wet, boggy ground.
Saline Contains salt (usually in water).
Saltmarsh A coastal habitat of grassland regularly flooded by seawater.
Water meadows Riverside meadows that are regularly flooded.
Weald Generally a heavily wooded area, from the Old English for 'forest'.

Photo credits

Bloomsbury Publishing would like to thank the following for providing photographs and for permission to reproduce copyright material within this book. While every effort has been made to trace and acknowledge all copyright holders, we would like to apologise for any errors or omissions, and invite readers to inform us so that corrections can be made to future editions.
Key to page positions: T = top; L = left; R = right; B = bottom.

5 Nigel J. Harris/Shutterstock; **8** Vic Thornley/Shutterstock; **12** Aerial Essex/Getty Images; **14** Caroline Legg via Flickr; **16** Amy Robjohns; **20** SanderMeertinsPhotography/Shutterstock; **24** Josephine Snell; **26** Matt Phelps; **28** Rob Newman; **29** John Gillett/Shutterstock; **32** Matt Phelps; **34** Birol Dincer/Shutterstock; **41** Stephan Morris/Shutterstock; **46** Lucy Boynton/iStock; **47** Dave Brassington; **48** Wirestock Creators/Shutterstock; **49L** Green and Blues/Shutterstock; **49R** Paul Abrahams/Shutterstock; **60** Matt Phelps; **62** Iliuta Goean/Shutterstock; **64** Malcolm Hunt/RSPB Images; **67** Neil Owen; **72** Matt Phelps; **74** Sandra Standbridge/Getty Images; **80** Dave Brassington; **82** Matt Phelps; **85** Dan Owen; **92** David Campbell; **94** Marc Read; **95** Marc Read; **102** Matt Phelps; **107** Paul Sawer/RSPB Images; **109** Peter Steward; **110** David Brassington; **118** Denja1/iStock; **119T** Andrew M. Allport/Shutterstock; **119B** Roger Meerts/Shutterstock; **122** Sandra Standbridge/Getty Images; **123** Matt Phelps; **127** Alex Cooper Photography/Shutterstock; **128** David Campbell; **129R** Agami Photo Agency/Shutterstock; **132** Erni/Shutterstock; **133** Dave Brassington; **134** Andy Reagon and Chrissy McClarren/Wikimedia; **147** Matt Phelps; **148** Erni/Shutterstock; **150B** Dave Brassington; **156** Nigel Voaden; **162** Nickos/Getty Images; **163** David Campbell; **164** Marco Rolleman/Shutterstock; **166** Matt Phelps; **167** Matt Phelps; **168** Matt Beauchamp/Shutterstock **170** Fascinadora/iStock **174** Bouke Atema/Shutterstock **175T** Amy Robjohns; **175B** Erni/Shutterstock; **176** Mark Christopher Cooper/Shutterstock **179** Matt Gibson/Shutterstock **180** David Campbell **183** Jeremy Halls via Flickr.

All other photos are the author's own.

About the author

Ed Stubbs is the Deputy Editor of *Birdwatch* magazine and BirdGuides. Based in Surrey, birding has been part of his life since he was a child, with a passion for patch birding a constant throughout. He believes that endless excitement and learning can be had by birding close to home. He enjoys sharing his knowledge, experience and advice to encourage others to embrace patch and local birding.

Index

Page numbers in italics refer to illustrations.